THE PENTAGON BUILDING PERFORMANCE REPORT

January 2003

1801 Alexander Bell Drive, Reston, Virginia 20191

ABSTRACT

Following the September 11 crash at the Pentagon of an airliner commandeered by terrorists, the American Society of Civil Engineers established a building performance study team to examine the damaged structure and make recommendations for the future. The members of the team reviewed available information on the structure and the crash loading and drew on focused assessments by others. In addition to analyzing the essential features of column response to impact, they investigated the residual frame capacity and the structural response to the fire. Plausible mechanisms for the response of the structure to the crash were established. Recommendations are offered for future design and construction along with suggestions for research and development.

Library of Congress Cataloging-in-Publication Data

The Pentagon building performance report.
p. cm.
Includes bibliographical references.
ISBN 0-7844-0638-3
1. Buildings—Blast effects.
2. Building, Bombproof.
3. Pentagon (Va.)—Evaluation.
I. American Society of Civil Engineers.

TH1097 .P46 2002
693.8'54—dc21 2002074554

Cover Photograph: AFP PHOTO/SHAWN THEW

THE PENTAGON BUILDING PERFORMANCE REPORT AUTHORS

Paul F. Mlakar, Ph.D., P.E., Lead
Technical Director
U.S. Army Corps of Engineers

Donald O. Dusenberry, P.E.
Principal
Simpson Gumpertz & Heger, Inc.

James R. Harris, Ph.D., P.E.
Principal
J.R. Harris & Company

Gerald Haynes, P.E.
Fire Protection Engineer
Bureau of Alcohol, Tobacco, and Firearms

Long T. Phan, Ph.D., P.E.
Research Structural Engineer
National Institute of Standards and Technology

Mete A. Sozen, Ph.D., S.E.
Kettelhut Distinguished Professor of Structural Engineering
Purdue University

TABLE OF CONTENTS

EXECUTIVE SUMMARY

The Pentagon, the headquarters of the United States Department of Defense, was constructed between September 1941 and January 1943. A major renovation of the entire 6.6 million sq ft facility began in 1999 and is scheduled for completion in 2010. On September 11, 2001, a hijacked commercial airliner was intentionally crashed into the building in an act of terrorism. One hundred eighty-nine persons were killed and a portion of the building was damaged by the associated impact, deflagration, and fire. That same day the American Society of Civil Engineers established a building performance study (BPS) team to examine the damaged structure and make recommendations for the future. Team members possess expertise in structural, fire, and forensic engineering.

The BPS team's analysis of the Pentagon and the damage resulting from the attack was conducted between September 2001 and April 2002. The members of the BPS team inspected the site as soon as was possible without interfering with the rescue and recovery operations. They reviewed the original plans, the renovation plans, and available information on the material properties of the structure. They scrutinized aircraft data, eyewitness information, and fatality records; consulted with the urban search and rescue engineers, the chief renovation engineer, and the engineer in charge of the crash site reconstruction; and examined the quick, focused assessments of the disaster conducted by the United States Army Corps of Engineers and Pentagon Renovation Program staff.

On the basis of this information the BPS team analyzed the essential aspects of the response of the structural system of the Pentagon to the crash. Impact analyses revealed that the spirally reinforced columns could withstand substantial dynamic lateral loads and deflections. Static analyses indicated that the floor system was capable of significant load redistribution without collapse when several adjacent supporting columns were removed or severely damaged by an extreme action. Thermal analyses showed that the ensuing fire could have sufficiently weakened some damaged frame members to result in collapse within 20 minutes of initiation.

The BPS team concluded that the impact of the aircraft destroyed or significantly impaired approximately 50 structural columns. The ensuing fire weakened a number of other structural elements. However, only a relatively small segment of the affected structure collapsed, approximately 20 minutes after impact. The collapse, fatalities, and damage were mitigated by the Pentagon's resilient structural system. Very few upgraded windows installed during the renovation broke during the impact and deflagration of aircraft fuel.

The BPS team recommends that the features of the Pentagon's design that contributed to its resiliency in the crash—that is, continuity, redundancy, and energy-absorbing capacity—be incorporated in the future into the designs of buildings and other structures in which resistance to progressive collapse is deemed important. The team further advocates that additional research and development be conducted in the practical implementation of measures to mitigate progressive collapse and in the deformation capacity of spirally reinforced columns subjected to lateral loads applied over the height of the column.

1. INTRODUCTION

1.1 SPONSOR AND PURPOSE

On the morning of September 11, 2001, as part of a terrorist action involving four hijacked aircraft, a commercial airliner was crashed into the Pentagon. That afternoon the American Society of Civil Engineers established a building performance study (BPS) team of volunteers to examine the structural performance of the building in this catastrophe. This study follows a similar examination of the April 19, 1995, bombing of the Murrah Federal Office Building, in Oklahoma City, and parallels a study of the September 11 World Trade Center terrorist attack.

The purpose of the study was to examine the performance of the structure in the crash and the subsequent fire for the benefit of the building professions and the public. This does not imply that buildings should be expected to survive such events. However, this examination of the Pentagon reveals some useful information about the ability of structures to survive extreme forces.

In fact, the Pentagon structure survived this extraordinary event better than would be expected. Observations comparing and contrasting the construction of the Pentagon to current standards are made where they are pertinent to the observed behavior. Recommendations are also made for studies that could lead to an increased understanding of such phenomena.

Pentagon Renovation Program Office

Figure 1.1 The Pentagon on September 5, 2001

1.2 STUDY TEAM

The BPS team included specialists in structural, fire, and forensic engineering. The following six individuals constituted the core group and are the authors of this report:

Paul F. Mlakar, Ph.D., P.E., Lead
Technical Director
U.S. Army Corps of Engineers
Vicksburg, Mississippi
Specialty: blast-resistant design; investigator, Murrah Federal Office Building study

Donald O. Dusenberry, P.E.
Principal
Simpson Gumpertz & Heger, Inc.
Waltham, Massachusetts
Specialty: blast effects and structural design

James R. Harris, Ph.D., P.E.
Principal
J.R. Harris & Company
Denver, Colorado
Specialty: structural engineering

Gerald Haynes, P.E.
Fire Protection Engineer
Bureau of Alcohol, Tobacco, and Firearms
Washington, D.C.
Specialty: fire protection

Long T. Phan, Ph.D., P.E.
Research Structural Engineer
National Institute of Standards and Technology
Gaithersburg, Maryland
Specialty: concrete structural and fire engineering

Mete A. Sozen, Ph.D., S.E.
Kettelhut Distinguished Professor of Structural Engineering
Purdue University
Lafayette, Indiana
Specialty: behavior of reinforced-concrete structures

Many professional colleagues of these team members voluntarily contributed to the work of this team, in much the same way that the public assisted the victims of the disaster. These individuals are acknowledged in appendix A. W. Gene Corley, the BPS team leader for the World Trade Center Study, facilitated cooperation between the two study teams.

1.3 THE PENTAGON

The Pentagon is one of the largest office buildings in the world, encompassing about 6.6 million sq ft of floor space. Its name refers to the five sides of the building (figure 1.1), but the Pentagon is also five stories high and is subdivided into five circumferential rings. In the upper three stories, the rings are separated by light wells; the second well from the interior extends to the ground over most of its length and serves as an interior driveway. Ten radial corridors provide connection from ring to ring.

As the Pentagon approached its 50th anniversary in service, an extensive renovation was planned. (See "The Pentagon Project," *Civil Engineering*, June 2001.) The actual construction began in 1999, and by September 11, 2001, one-fifth of the renovation was essentially complete. Structurally the renovation was not major; the most significant changes were the addition of new elevators, stairs, escalators, and mechanical equipment rooms. Additionally, the exterior walls and windows were upgraded to provide a measure of resistance to extreme pressures.

Associated Press

Figure 1.2 Aircraft impact

DoD photo by Tech. Sgt. Cedric H. Rudisill

Figure 1.3 Crash damage

1.4 THE CRASH

At 9:38 A.M. on September 11 an airliner was flown into the first story of the Pentagon. The impact occurred in the renovated portion of the building approximately 140 ft to the south of the boundary between the renovated section and the next section scheduled to be renovated. (Figure 1.2, a photograph taken by a security camera, shows the plane impacting the building at ground level.) The aircraft sliced through the building into the section not yet renovated. The impact and the fire initiated by the fuel in the airplane that immediately spread widely in the structure took the lives of all 64 people aboard the aircraft and 125 occupants of the Pentagon.

1.5 EXTENT OF DAMAGE

Figure 1.3 presents an exterior view of the extent of damage from the crash, including a collapsed portion of Ring E at the point of impact, beyond which the impact destruction from the decelerating aircraft continues; the subsequent devastation from the fire is also evident. The superior performance of the improved window system incorporated during the renovation is evident on the right.

1.6 SCOPE OF REPORT

Two issues of structural performance commanded attention in this study. First, the collapse that did occur was not immediate; this calls for an examination of the interaction between fire and structural performance. Second, many of the first-story columns in a portion of the structure that remained standing were destroyed during the crash. Such performance is desirable, and the reasons for it are of interest to the engineering profession.

Sections 2 through 6 of this report first detail the observations made directly by the team and review pertinent information from other sources that was significant in defining the performance. In section 7, approximate analyses of impact, structural, and fire behavior made by the team to illustrate the essential aspects of the performance are described. Sections 8 and 9 present findings and recommendations that are consistent with the purpose of the study.

2. REVIEW OF BUILDING INFORMATION

Volumes of information exist regarding the original construction of the Pentagon and its current renovation. The Pentagon Renovation Program Office assisted the BPS team in accessing the essential data for the purpose of this study.

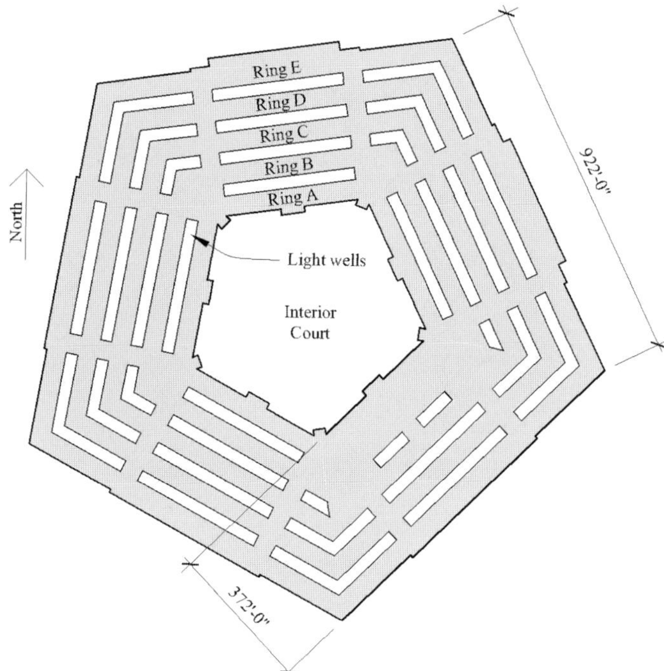

Figure 2.1 Overall plan of the Pentagon at the upper stories

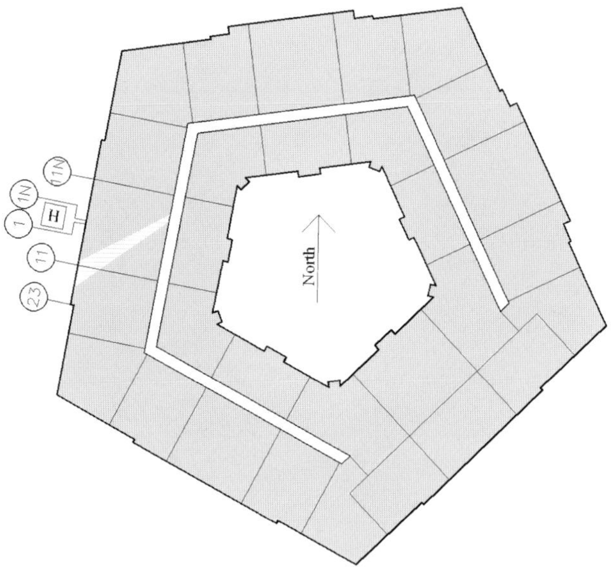

Figure 2.2 Overall plan at first story, showing expansion joints; hatching shows the aircraft impact zone; H is the helipad.

2.1 DOCUMENTS FOR ORIGINAL CONSTRUCTION

Figure 2.1 is a plan view of the five-sided building at the upper three stories, emphasizing the light wells between the rings and the radial corridors. Figure 2.2 is a plan view at the first story, with the structural expansion joints added. Note that the light well between rings B and C extends to the ground over most of the circumference. It is known as AE Drive. For future reference, figure 2.2 also shows the area impacted by the crash and a few key grid lines.

The original drawings for the Pentagon subdivide the building into "sections" A through E and "areas" 1 through 10 (figure 2.3). The drawings referring to areas use the U.S. style of labeling stories—the ground level being called the first floor—whereas the drawings referring to sections use the European style, the level above the ground level being called the first floor. The radial corridors are numbered to correspond to the areas.

The Pentagon is in the midst of a major renovation program, and the work is phased in five "wedges" that do not correspond to either the sections or the areas. Each wedge is centered on a building vertex and consists of the portion of the building between the midpoint of adjacent sides. The renovation of Wedge 1 began in 1999 and was essentially complete at the time of the crash.

The original structural system, including the roof, was entirely cast-in-place reinforced concrete using normal-weight aggregate. Most of the structure used a specified concrete strength of 2,500 psi and intermediate-grade reinforcing steel (yield of 40,000 psi). The floors are constructed as a slab, beam, and girder system supported on columns, most of which are square. Figures 2.4 through 2.8 define the typical framing. Member sizes vary with framing arrangements and special loads. The area of interest in this study was populated by the typical members shown in the figures. The column sizes vary in each story—generally from about 21 by 21 in. in the first story to 14 by 14 in. in the fifth story—but there are many exceptions. Nearly all the columns that support more than one level are spirally reinforced. The remaining columns have ties. The floor spans are relatively short by modern standards: 5.5 in. slabs span to 14 by 20 in. beams at 10 ft on center. The typical beam spans are 10 or 20 ft, with some at 15 ft. Girders measuring 14 by 26 in. span 20 ft parallel to the exterior walls and support a beam at midspan.

Figure 2.9 is a cross section transverse to the exterior wall. The roof at Ring E is gabled (as are those over Ring A and the radial corridors). Slabs 4.5 in. thick span perpendicular to the exterior wall with spans varying from about 8 to 11 ft. The slabs are supported by 12 by 16 in. purlins that span to rafter frames, which are 20 ft on center. The rafters are generally 16 by 24 in. and align with the floor beams and columns below. In general, the purlins do not align with the floor girders and columns below.

The roof over rings B, C, and D consists of a nearly flat pan joist and slab system. The joist stems are 6 in. wide by 8 in. deep and the slab is 2.75 in. thick. The joists are 26 in. on center and span 20 ft. The roof over the corridors is 4.5 in. thick and spans 10 ft. The joists and slab are supported by 14 by 20 in. girders that are in line with the floor girders.

The perimeter exterior walls of Ring E are faced in limestone and backed with unreinforced brick infilled in the concrete frame. Nearly all remaining exterior walls are 10 in. concrete. The first story at AE Drive is brick infilled in the concrete frame, with no windows. The concrete walls have 5 by 7 ft openings for windows and include columns built in as pilasters, corresponding to column locations below, and girders reinforced within the wall. Figure 2.10 is an elevation at a typical light well wall.

Slabs, beams, and girders all make use of straight and trussed bars. Except for the top reinforcement in the short spans adjacent to longer spans, there are no continuous top bars. However, approximately half of the bottom bars are made continuous by laps of 30 to 40 bar diameters at the supports. Beams and girders typically have open-topped stirrups. The longer spans generally have approximately equal areas of steel at the critical sections.

Any building is a product of its times. The Pentagon was constructed between September 1941 and January 1943. At that time the national standard predominantly used for reinforced-concrete buildings was ACI 501-36 as developed by the American Concrete Institute (ACI). Although no reference to ACI 501-36 was found in the drawings, it is very likely that this code affected decisions about member sizing and proportioning for the Pentagon structure. A brief review of some of its basic requirements is in order.

ACI 501-36 was based on working stress design. The allowable stress for the intermediate-grade billet steel used in the Pentagon was 20,000 psi. For the design concrete strength of 2,500 psi, the allowable unit shear stress for beams with properly

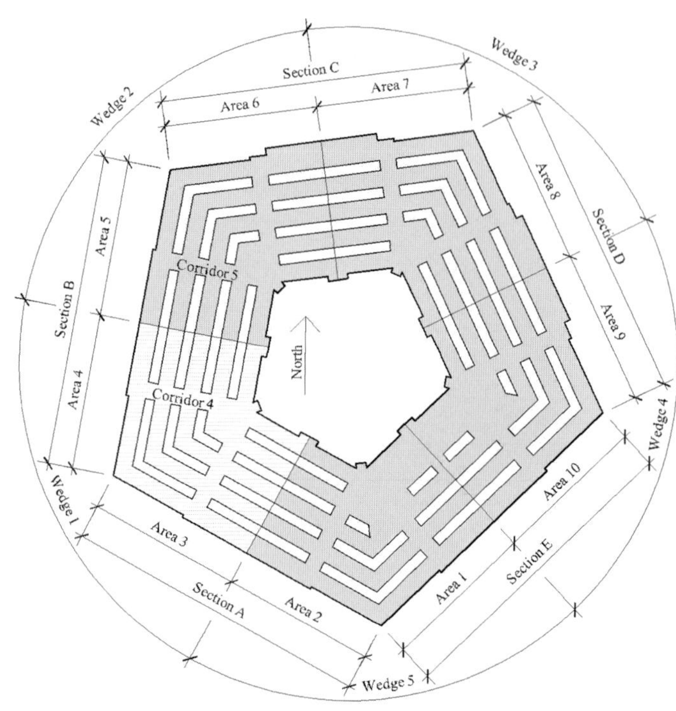

Figure 2.3 Designation of areas and sections used in the original construction and of wedges used in the ongoing renovation; Wedge 1 is hatched.

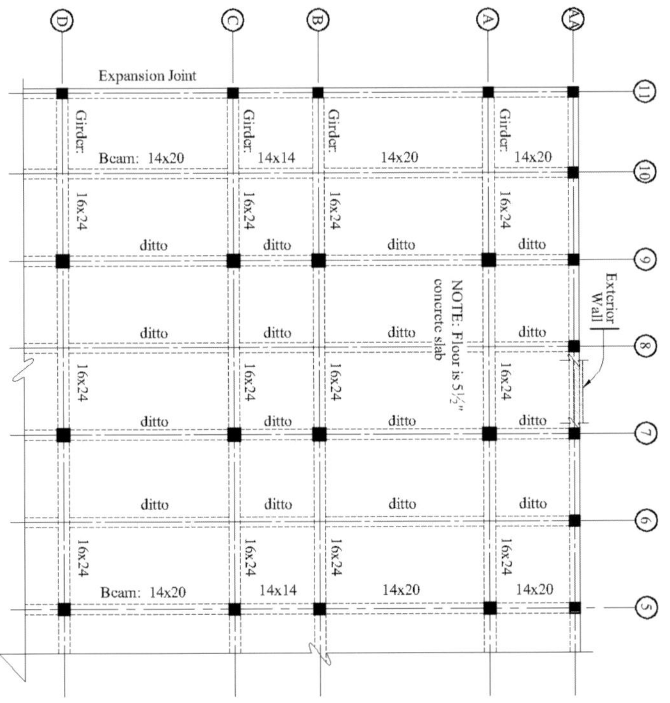

Figure 2.4 Partial plan showing typical slab, beam, and girder framing of floors

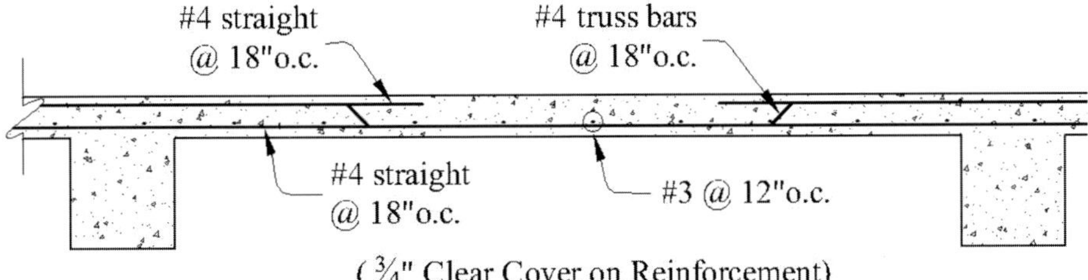

Figure 2.5 Detail of typical floor slab

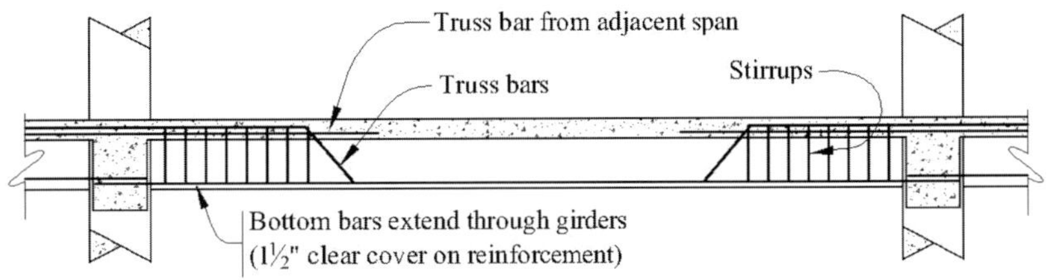

Figure 2.6 Detail of typical beam

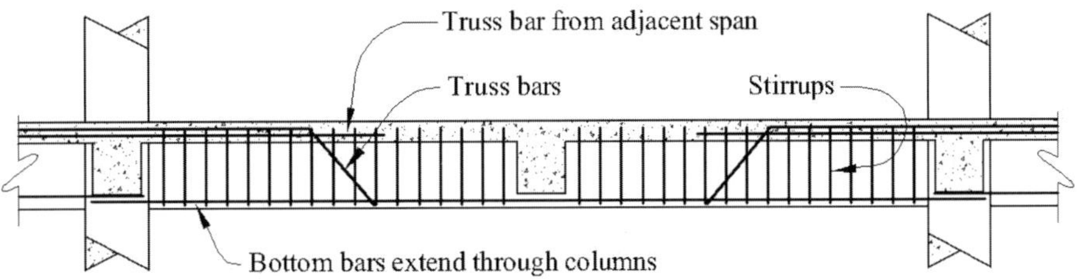

Figure 2.7 Detail of typical girder

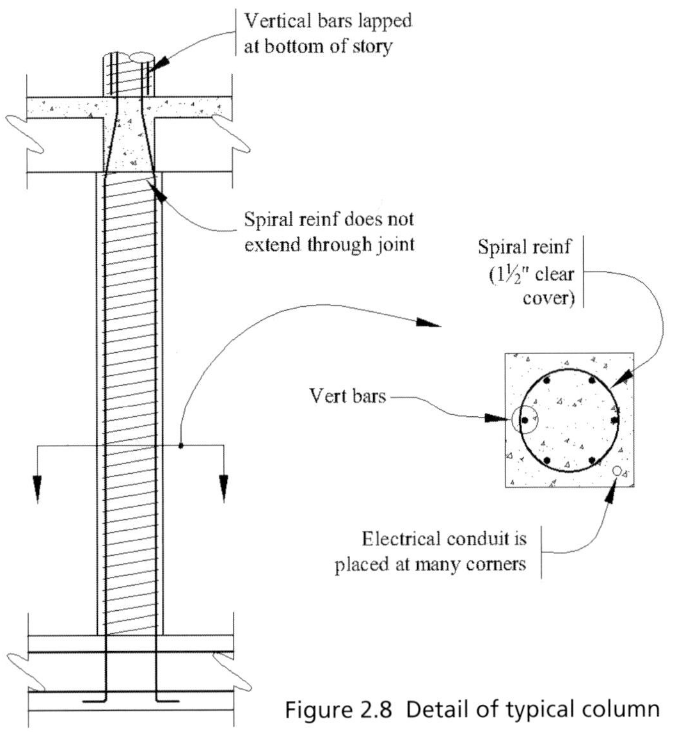

Figure 2.8 Detail of typical column

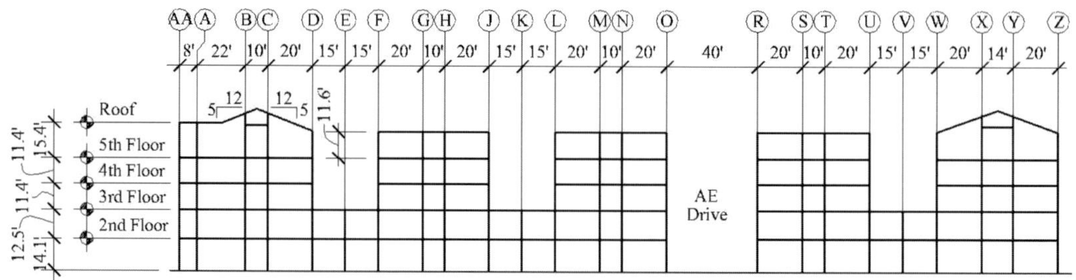

Figure 2.9 Transverse cross section through the rings

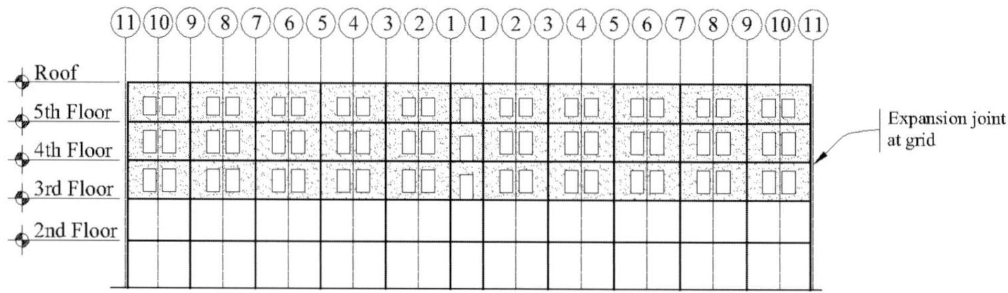

Figure 2.10 Elevation of light well wall

designed web reinforcement was 150 psi. The unit bond stress for deformed bars was 125 psi for the same strength of concrete.

The axial load, P, permitted on reinforced-concrete columns with spiral reinforcement was defined by the expression

$$P/A_g = 0.225 f'_c + \rho_g \cdot f_s$$

where

A_g =gross cross-sectional area;

f'_c=compressive strength of concrete (6 x 12 in. cylinder);

ρ_g = ratio of longitudinal reinforcement area to area of cross section;

f_s = permissible unit stress (16,000 psi for intermediate-grade steel).

This expression was conservative. The estimated service load of the column was set at approximately one-third of its expected strength on the basis of the specified compressive strength of the concrete and the specified yield stress of the reinforcement.

At that time continuity in reinforced concrete was still difficult to handle analytically. Thus ACI 501-36 specified the following moment coefficients to be used in design.

Negative moment at face of first interior support: $(1/10)wl^2$

Negative moment at face of interior supports: $(1/12)wl^2$

Positive moment at center of exterior spans: $(1/10)wl^2$

Positive moment at center of interior spans: $(1/12)wl^2$

It is especially interesting to note that the interior-span positive and negative moments were to be of the same magnitude.

The physical characteristics of the Pentagon structure suggest that its design may have been influenced strongly by the book *Reinforced Concrete Construction* (Hool and Pulver 1937). This work describes a floor system quite similar to that depicted in figure 2.4. It is also of interest to note that this reference recommends a reinforcing arrangement similar to that in figure 2.7 for the girders. A critical attribute of the Pentagon structure was the continuity of at least half of the bottom reinforcement across the column line to lap for a distance of at least 30 bar diameters.

Table 2.1 Column Type 14 Load and Capacity

Story	Dead (kip)	Live (kip)	Total (kip)	Allowable (kip)	Ratio
5	37	8	43	119	2.75
4	77	51	128	150	1.18
3	116	77	192	203	1.06
2	157	113	170	283	1.05
1	200	149	349	351	1.01

The columns were very important to the overall performance of the Pentagon; thus their original design was examined somewhat more closely. Table 2.1 shows that one of the most typical columns (type 14) apparently was designed to be economical by the original designers, because the margin of allowable capacity to demand was very close to unity in the lower stories. This computation ignores any bending moment from lateral loads, which at the time of the design were probably shown to be accommodated by the one-third increase in allowable stresses that was the fashion at the time. The live loads were reduced by 20 percent for columns supporting more than one floor, the common rule for storage loads.

Examination of the column design data leads to the conclusion that the minimum size used for columns was 14 in. square and that tied reinforcement was used until higher loads demanded a change. The first change was to spiral reinforcement, a 25 percent increase in allowable load by the standard of the day. The next change was in the size of the column. Given the nature of formwork at the time, today's imperative of keeping column sizes constant was obviously not an issue.

Given the attention paid by the planners of the Pentagon renovation to the capacities of the existing structure, the BPS team did not study the lateral load capacity in much detail. A factor of note is that the seismic demands by current standards are less at the Pentagon site than the demands used in the planning studies. Given the higher material strengths found since those studies, as well as some very simple check analyses, the Pentagon structure would probably be evaluated as having a seismic capacity in line with guidelines for existing buildings.

2.2 DOCUMENTS FOR RENOVATION

Planning for the renovation of the Pentagon began in the late 1980s as the building was approaching 50 years in service. The planning and execution of the renovation led to the creation of three documents that are useful in understanding the structure and, by extension, its performance.

The July 1993 *Structural Renovation Study,* conducted by a joint venture of Daniel, Mann, Johnson, and Mendenhall and 3DI for the Army Corps of Engineers, Baltimore District, was commissioned as part of the overall evaluation of the questions concerning the merit and the details of renovating the Pentagon. The report addressed the load capacity, the general condition, and the seismic resistance. The stated purpose was to examine the structural implications of new stairs, elevators, escalators, and duct shafts; the removal of steep floor-to-floor ramps; the creation of large, multistory spaces for an auditorium and multipurpose rooms; and the addition of more mezzanine space in the basement.

The general condition of the concrete structure was found to be good. This study was completed before testing of materials had been completed, and there was uncertainty about the strength of the reinforcing steel. Many of the calculations, performed in 1991 and 1992, assumed structural grade rebar, f_y = 33 ksi. Before the report was issued, it had been concluded that intermediate grade (f_y = 40 ksi) was more appropriate. Concrete strength was assumed to have been the specified 2,500 psi. The report stated that some pile load tests had been performed in the early 1970s, and those tests showed an allowable capacity nearly twice the 30 tons per pile required on the original drawings.

The floor live-load capacity was confirmed at 150 psf, with no reductions based upon area. This corroborated the June 25, 1944, report, *Pentagon Project,* which indicated that this live load was used because it was anticipated that the building would be used for record storage following the war. Capacities of typical members were checked using the 1989 edition of ACI 318, *Standard Building Code for Reinforced Concrete*; a few comparisons with the 1941 edition were made to see if there was anything about the original design that was not conservative in comparison with current standards. Reference is made to another report, *Pentagon Renovation Program Standards and Criteria,* for loads to use in the design for the renovation.

The *Structural Renovation Study* included a brief examination of the lateral-load-resisting system of the building. Although the report concluded that the military design manuals did not require a seismic evaluation and upgrade, seismic loads dominated this examination. A typical frame in the radial direction was evaluated using 80 percent of the seismic loading required for new buildings, which was the standard method at the time. Three different standards for new buildings were considered: one based upon the 1975 edition of the Structural Engineers Association of California (SEAOC) *Blue Book,* a second upon the 1990 edition of the same document, and a third upon the 1990 edition of the Building Officials and Code Administrators' (BOCA) *National Building Code.* A fourth analysis for wind load from the 1992 supplement to the BOCA code was also performed. The loading was largest in the first method cited and decreased successively in each of the others.

The basic finding was that the capacity of all columns except the fifth-story columns exceeded the demands of the required load, while the capacity of the beams was slightly less than the demand for the largest of the loads. The computations were performed using 33 ksi for the yield strength of the reinforcing; substituting 40 ksi shows that these capacities would be adequate.

From the analysis it was also found that the larger loads would cause pounding of adjacent structures at the expansion joints. It was postulated that shear walls or braces would be required to resolve the deficiencies and that the natural concentration of forces resulting from such strengthening would inevitably overload the foundations, which would then also require additional capacity. Concern was also expressed about the lack of full-length top reinforcement in the beams and girders. The conclusion was that seismic upgrading was not necessary, primarily because of the low hazard for strong ground motions.

On the whole this study clearly indicates that the reinforced-concrete structure was competently designed and built for heavy loads, was showing very little distress, and could be relied upon for continued good and safe service.

The March 1998 *Renovation of the Pentagon: Basis of Design: Ready-to-Advertise Submission: Wedge 1,* produced by Hayes, Seay, Mattern & Mattern for the Army Corps of Engineers, Baltimore District, describes the criteria used for the design of the renovation, focusing on Wedge 1 of the Pentagon. The scope of the structural work (in all wedges) includes the construction of 17 new three-story mechanical rooms to be built in portions of the existing light wells; the installation of several new elevators, escalators, and duct shafts; the replacement of interfloor ramps with level floors; the construction of new utility tunnels and the renovation of others;

some construction over AE Drive; the inclusion of new doors in the light wells; the closure of some existing floor openings; and the installation of new full-height stairs.

The structural design criteria, in general, were those contained in the current editions of familiar standards for the design of concrete, steel, and masonry structures. The live loads were generally those required by the standard ASCE 7, *Minimum Design Loads for Buildings and Other Structures* (1993 edition), with the exception of general office areas, which were to be designed for an 80 psf live load plus a 20 psf allowance for partitions. Wind and snow loads were also to be as specified in ASCE 7, and seismic loads were as defined in military Technical Manual 5-809-10, which is an application of the SEAOC *Blue Book*—hence it would be very similar to the 1994 *Uniform Building Code*.

The new construction consisted primarily of either filling in existing voids in the original construction or creating framing around new openings. This new work mainly entailed concrete slabs on composite steel deck and beams. Where the loads exceeded the capacity of the existing concrete columns, new steel columns and foundations were provided.

The structural class for fire resistance was type 1B (protected, noncombustible) as defined in the BOCA *National Building Code*. The original concrete structure inherently possessed such ratings, and the new steel beams and columns achieved the protection with sprayed-on insulation (generally two-hour rating for floors and members supporting only one level and three-hour rating for members supporting more than one level).

As described previously, the original exterior Ring E wall is mostly non-load-bearing masonry infilled in a concrete frame. The exterior surface is 5 in. thick limestone, which covers the frame, backed by 8 in. unreinforced brick that is infilled in the frame. In some areas the backing is a cast-in-place concrete wall. At the locations inspected for this study, the brick infill at the fifth story was not in contact with the columns, but was separated by a 2 in. gap crossed by metal ties between the mortar joints and dovetail slots in the column face. (According to consultation with an engineer from the authoring firm, at other locations the brick was mortared tight to the columns.) Concrete columns exist at 20 ft on center in the fifth story and at 10 ft on center in the lower stories. The fifth story has no windows, and the brick is interrupted by a concrete beam between the fifth floor and the eave of the roof. The remaining stories have 5 by 7 ft windows in the majority of the 10 ft bays, and the head of the window is the soffit of the concrete edge beam of the floor above.

3. REVIEW OF CRASH INFORMATION

The volume of information concerning the aircraft crash into the Pentagon on September 11 is rather limited. Through the cooperation of transportation, law enforcement, and news organizations the BPS team was able to collect the essential data for the purpose of this study.

3.1 AIRCRAFT DATA

The impacting airplane was a Boeing 757-200 aircraft, originally delivered in 1991. This aircraft was designed to accommodate approximately 200 passengers and 1,670 cu ft of cargo. The wingspan, overall length, and tail height (see figure 3.1) were respectively 124 ft 10 in., 155 ft 3 in., and 44 ft 6 in. Maximum takeoff weight was 255,000 lb, including up to 11,275 gal of fuel. Much of the aircraft fuel was contained in wing tanks. The aircraft was designed to cruise up to 3,900 nautical miles at a speed of Mach 0.80 (approximately 890 ft/s). The two engines were manufactured by Rolls-Royce and had 44,000 lb of combined thrust.

When the aircraft departed from Washington's Dulles International Airport on the morning of September 11, 2001, it held 64 persons—passengers and crew members—and enough fuel for the cross-country trip to Los Angeles. According to the National Transportation Safety Board, the aircraft weighed approximately 181,520 lb and was traveling at 460 knots (780 ft/s) on a magnetic bearing of 70 degrees when it struck the Pentagon. The aircraft had on board approximately 36,200 lb (5,300 gal) of fuel at the time of impact.

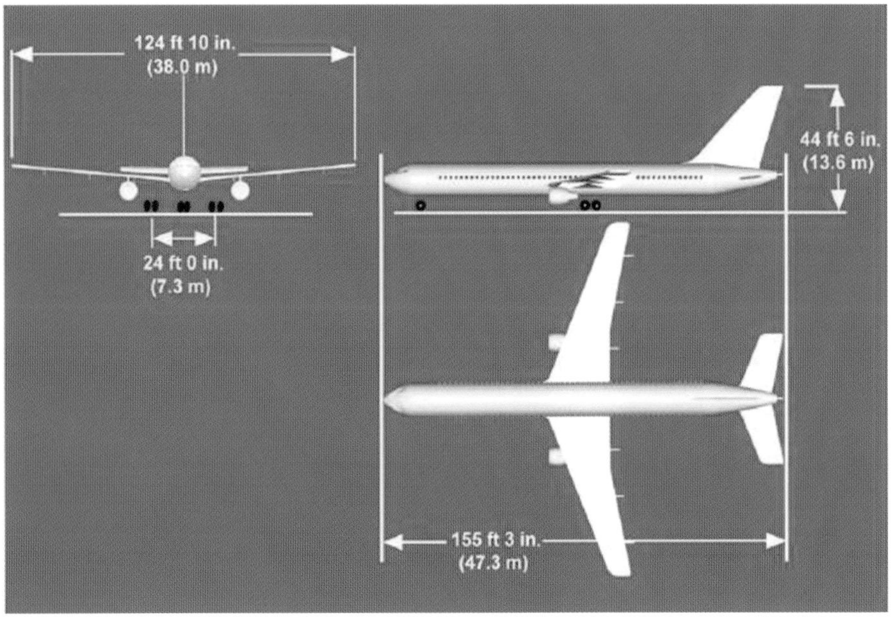

Figure 3.1 Dimensions of Boeing 757-200 aircraft

According to Boeing engineers, the weight in each wing was composed of the following:

Exposed wing structure: 13,500 lb
Engine and struts: 11,900 lb
Landing gear: 3,800 lb
Fuel: 14,600 lb
Total: 43,800 lb

The balance of the weight was in the fuselage. In the normal course of use the center fuel tank is the last filled and the first used. Thus the weight of the fuselage at the time of impact was 181,520 – (2 x 43,800) = 93,920 lb. Of this, 36,200 – (2 x 14,600) = 7,000 lb was fuel in the center tank.

3.2 EYEWITNESS INTERVIEWS

On January 8, 2002, BPS team leader Paul Mlakar interviewed three eyewitnesses—two of whom witnessed the impact of the aircraft and one of whom witnessed the subsequent partial collapse of the building. All three are professional staff members of the Pentagon Renovation Program Office and collectively provide a coherent and credible account of the events.

Frank Probst, 58, is a West Point graduate, decorated Vietnam veteran, and retired army lieutenant colonel who has worked for the Pentagon Renovation Program Office on information management and telecommunications since 1995. At approximately 9:30 A.M. on September 11 he left the Wedge 1 construction site trailer, where he had been watching live television coverage of the second plane strike into the World Trade Center towers. He began walking to the Modular Office Compound, which is located beyond the extreme north end of the Pentagon North Parking Lot, for a meeting at 10 A.M. As he approached the heliport (figure

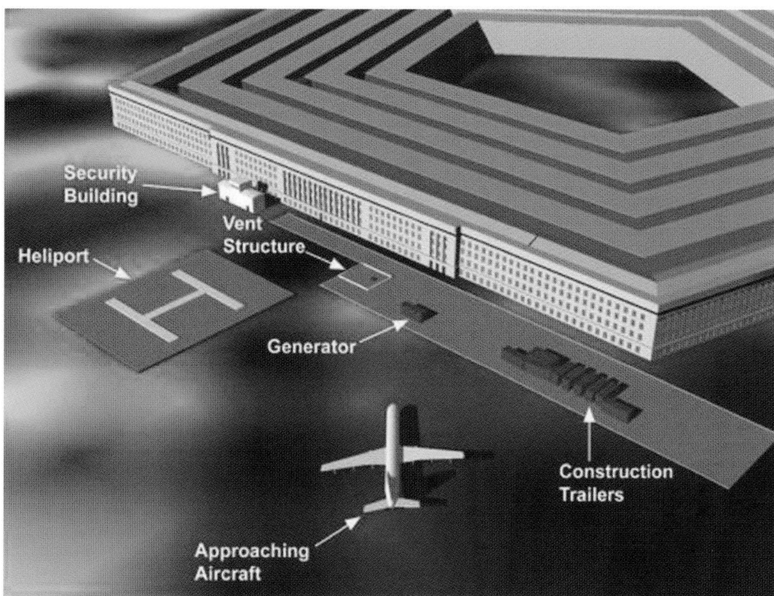

Figure 3.2 Pentagon and approaching aircraft, viewed from the southwest

3.2) he noticed a plane flying low over the Annex and heading right for him. According to the Arlington County after-action report (Arlington County, 2002), this occurred at 9:38 A.M. The aircraft pulled up, seemingly aiming for the first floor of the building, and leveled off. Probst hit the ground and observed the right wing tip pass through the portable 750 kW generator that provides backup power to Wedge 1. The right engine took out the chain-link fence and posts surrounding the generator. The left engine struck an external steam vault before the fuselage entered the building. As the fireball from the crash moved toward him, Probst ran toward the South Parking Lot and recalls falling down twice. Fine pieces of wing debris floated down about him. The diesel fuel for the portable generator ignited while he was running. He noted only fire and smoke within the building at the point of impact. Security personnel herded him and others to the south, and he did not witness the subsequent partial collapse of the building.

Don Mason, 62, is a communications specialist who retired from the United States Air Force after 25 years of service. He has worked for the Pentagon Renovation Program Office on information management and telecommunications since 1996. At the time of the crash he was stopped in traffic west of the building. The plane approached low, flying directly over him and possibly clipping the antenna of the vehicle immediately behind him, and struck three light poles between him and the building. He saw his colleague Frank Probst directly in the plane's path, and he witnessed a small explosion as the portable generator was struck by the right wing. The aircraft struck the building between the heliport fire station and the generator, its left wing slightly lower than its right wing. As the plane entered the building, he recalled seeing the tail of the plane. The fireball that erupted upon the plane's impact rose above the structure. Mason then noticed flames coming from the windows to the left of the point of impact and observed small pieces of the facade falling to the ground. Law enforcement personnel moved Mason's vehicle and other traffic on, and he did not witness the subsequent partial collapse of the building.

Rich Fitzharris, 52, is an electrical engineer and a former residential contractor. He has been the operations group chief of the Pentagon Renovation Program Office since 1996. He was in the Modular Office Compound at the time of the crash and rushed to the site on foot, arriving before the partial collapse. He recalls that the building—near the area of impact—was in flames, and he remembers seeing small pieces of debris, the largest of which might have been part of an engine shroud. He was at the heliport when a portion of the structure collapsed. The collapse initiated at the fifth floor along the building expansion joint, proceeded continuously, and was completed within a few seconds. According to the Arlington County after-action report, this occurred at 9:57 A.M., or 19 minutes after impact.

3.3 SECURITY CAMERA PHOTOGRAPHS

A Pentagon security camera located near the northwest corner of the building recorded the aircraft as it approached the building. Five photographs (figures 3.3 through 3.7), taken approximately one second apart, show the approaching aircraft and the ensuing fireball associated with the initial impact. The first photograph (figure 3.3) captured an image of the aircraft when it was approximately 320 ft (approximately 0.42 second) from impact with the west wall of the Pentagon. Two photographs (figures 3.3 and 3.7), when compared, seem to show that the top of the fuselage of the aircraft was no more than approximately 20 ft above the ground when the first photograph of this series was taken.

Associated Press, all

Figure 3.3 Aircraft approaching the Pentagon

Figure 3.4 Fireball within one second of impact (note security building in silhouette)

Figure 3.5 Fireball within two seconds of impact

Figure 3.6 Fireball within three seconds of impact

Figure 3.7 Fireball within four seconds of impact; the shadow of the smoke cloud visible on the ground provides a reference for determining the height of the aircraft.

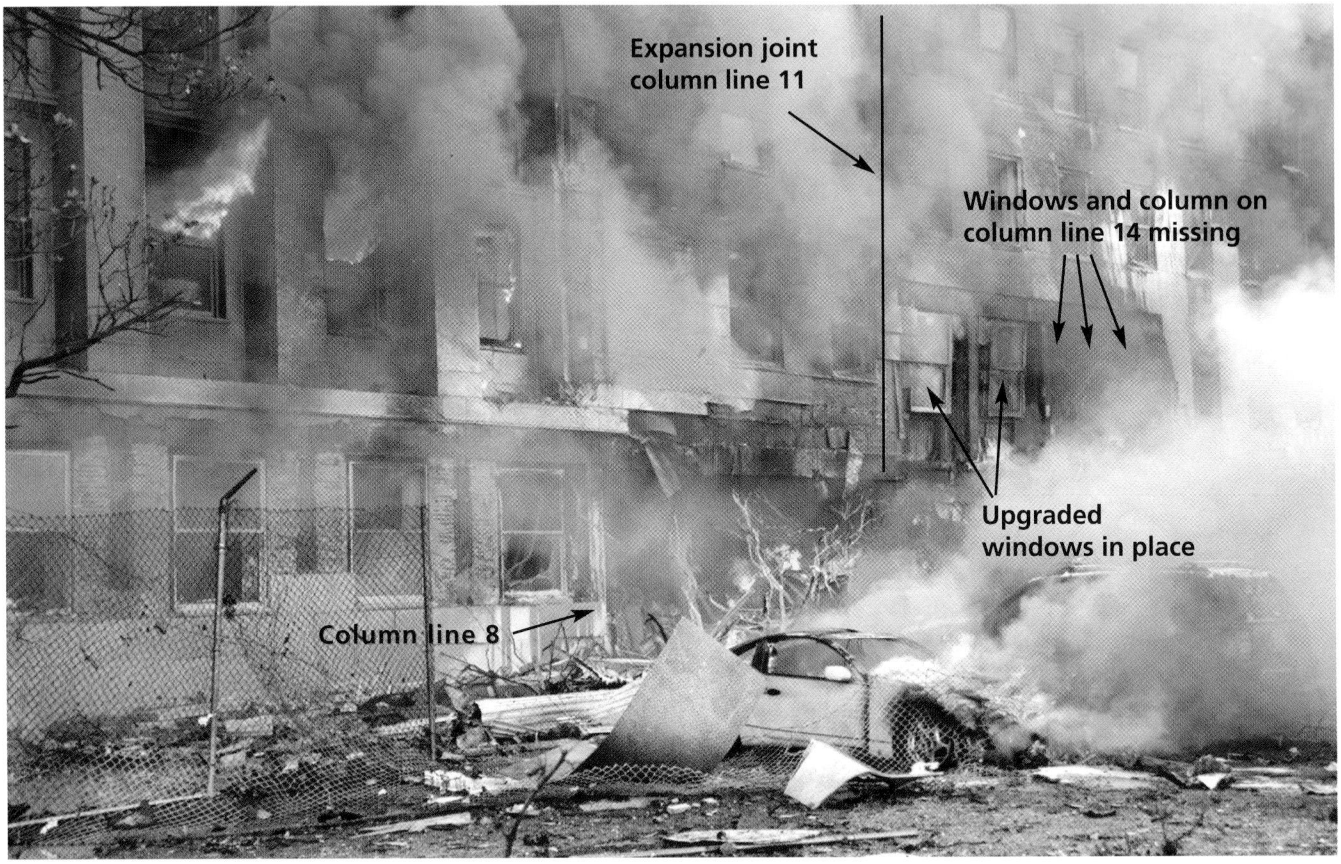

Associated Press/image enhancement by BPS team

Figure 3.8 Northern portion of impact area before collapse

Photograph by Steve Riskus/image enhancement by BPS team

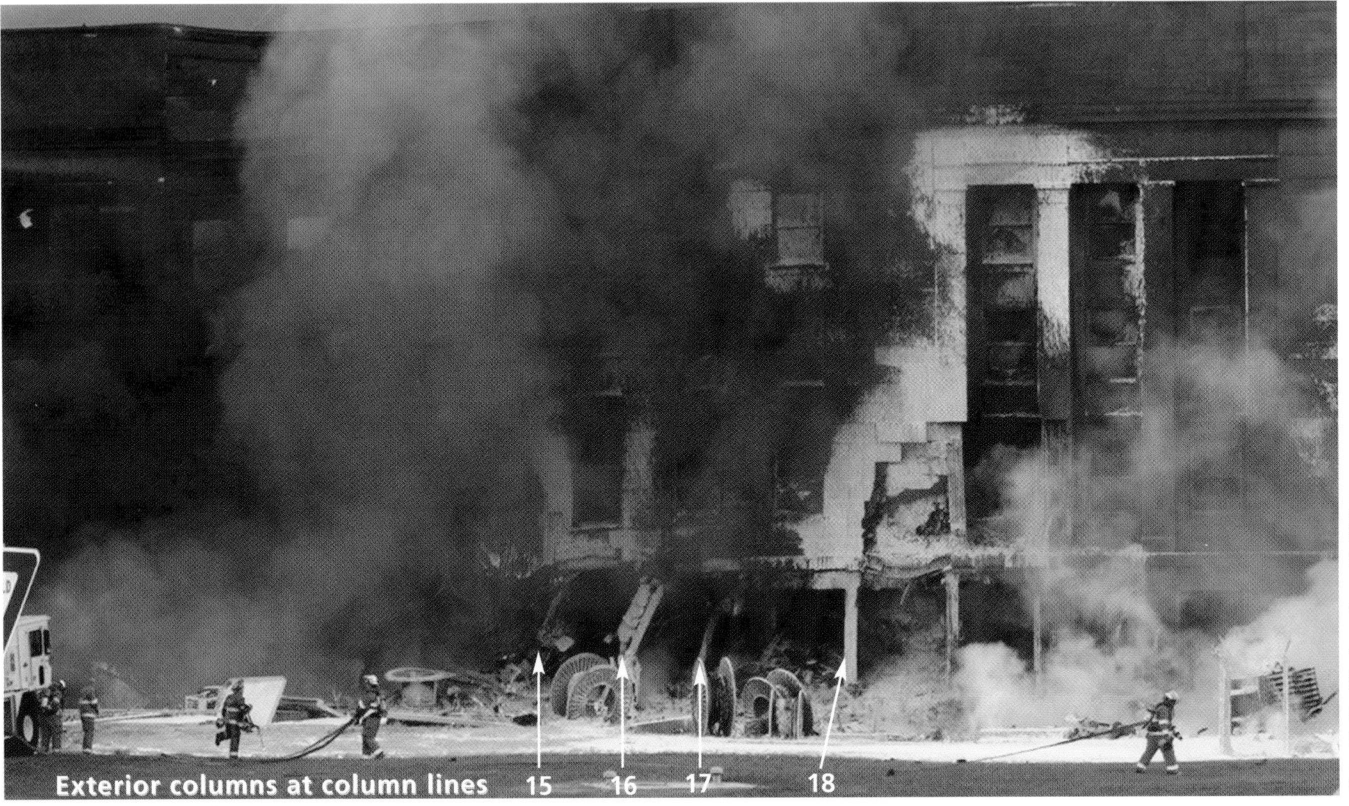

Figure 3.9 Southern portion of impact area before collapse

Figure 3.10 Impact location before collapse

3.4 PRECOLLAPSE PHOTOGRAPHS

A photograph (figure 3.8) taken by the Associated Press before the building to the south of the expansion joint collapsed provides useful information that the BPS team could not observe at the site. This photograph shows that the portion of the building that subsequently collapsed was displaced vertically by approximately 18 in. to 2 ft relative to the building north of the expansion joint. The facade was missing on the first floor as far north as column line 8 (the expansion joint is at column line 11), and on the second floor, the facade was missing between column lines 11 and 15. However, windows and their reinforcing frames were still in place between column lines 11 and 13 on the second floor.

The photograph also shows that the only column missing on the second floor in the west exterior wall of the building was at column line 14. The spandrel beam for the third floor and all third-floor exterior columns appears to be intact.

The photograph shows that upgraded windows installed as part of the Pentagon renovation were not broken by the impact or the fireball, even where the windows were located as close as 10 ft to the impact point of the fuselage.

A second photograph (figure 3.9) taken before the collapse reveals that first-floor exterior columns on column lines 15, 16, and 17 were severely distorted but still attached at least at their top ends to the second-floor framing. The vertical displacement noted previously also is evident in this photograph.

The precollapse state of the structure is illustrated for clarity in figure 3.10.

3.5 PHOTOGRAPH TAKEN DURING COLLAPSE

Figure 3.11 is a photograph taken at the moment of collapse, approximately 20 minutes after the aircraft struck the west wall of the Pentagon. The collapse extended to approximately column line 18 on the west face of the building.

DefenseLink

Figure 3.11 Portion of Ring E at moment of collapse

3.6 POSTCOLLAPSE PHOTOGRAPH

Figure 3.12—a photograph taken after the fire was extinguished, but before significant debris was removed from the collapse area or shoring was installed—shows the condition of the building after the collapse. The collapsed portion of Ring E extends from an expansion joint on column line 11 to approximately column line 18 on the west facade.

The limestone facade and brick dislodged from the framing, revealing the steel supports that had been added during the renovation. Window panes were generally still in place in windows in the area that collapsed. Scars from debris impact are visible to the south of the collapse area, as high on the facade as the tops of the fourth-floor windows.

3.7 SUMMARY OF THE IMPACT

The Boeing 757 approached the west wall of the Pentagon from the southwest at approximately 780 ft/s. As it approached the Pentagon site it was so low to the ground that it reportedly clipped an antenna on a vehicle on an adjacent road and severed light posts. When it was approximately 320 ft from the west wall of the building (0.42 second before impact), it was flying nearly level, only a few feet above the ground (figures 3.2 and 3.13, the latter an aerial photograph modified graphically to show the approaching aircraft). The aircraft flew over the grassy area next to the Pentagon until its right wing struck a piece of construction equipment that was approximately 100 to 110 ft from the face of the building (0.10 second before impact (figure 3.14). At that time the aircraft had rolled slightly to the left, its right wing elevated. After the plane had traveled approximately another 75 ft, the left engine struck the

DefenseLink

Figure 3.12 Ring E after collapse

Photograph, Air Survey Corporation; illustrated enhancement by BPS team

Figure 3.13 Aerial photograph modified to show approaching aircraft

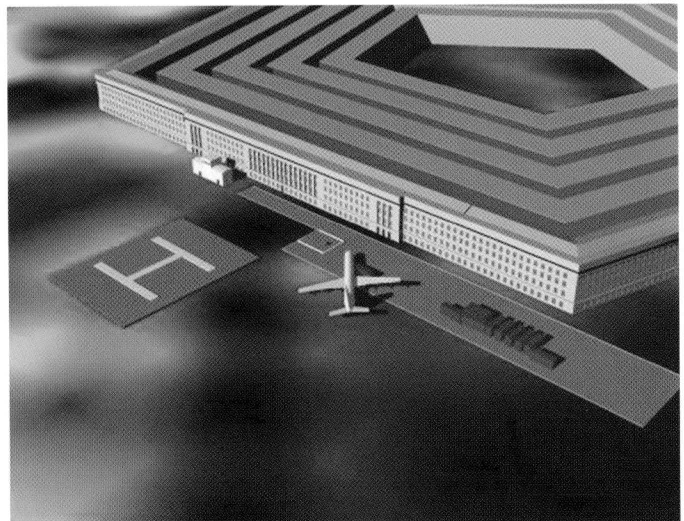

Figure 3.14 Aircraft at impact with generator

Figure 3.15 Aircraft at impact with vent structure

Figure 3.16 Aircraft at impact with the Pentagon

ground at nearly the same instant that the nose of the aircraft struck the west wall of the Pentagon (figure 3.15). Impact of the fuselage was at column line 14, at or slightly below the second-floor slab. The left wing passed below the second-floor slab, and the right wing crossed at a shallow angle from below the second-floor slab to above the second-floor slab (figure 3.16)

A large fireball engulfed the exterior of the building in the impact area. Interior fires began immediately.

The impact upon the west facade removed first-floor columns from column lines 10 to 14. First-floor exterior columns on column lines 9, 15, 16, and 17 were severely damaged, perhaps to the point of losing all capacity. The second-floor exterior column on column line 14 and its adjacent spandrel beams were destroyed or seriously damaged. Additionally, there was facade damage on both sides of the impact area, including damage as high as the fourth floor. However, in the area of the impact of the fuselage and the tail, severe impact damage did not extend above the third-floor slab.

Immediately upon impact, the Ring E structure deflected downward over the region from an expansion joint on column line 11 south to the west exterior column on column line 18 (figures 3.8–3.10). The deformation was the most severe at the expansion joint, where the deflection was approximately 18 in. to 2 ft.

The structure was able to maintain this deformed shape for approximately 20 minutes, at which point all five levels of Ring E collapsed from column line 11 to approximately column line 18 (figure 3.12).

3.8 FATALITY INFORMATION

Data provided by the Federal Bureau of Investigation included the locations where fatalities were found in the building. This information is reproduced as figures 3.17 and 3.18.

No fatalities from the aircraft were found on the second floor (figure 3.18). The figure shows no fatalities in the collapsed area above the first floor. In fact, it is likely that there were upper-floor fatalities in this area, but their remains were found on the first-floor level after the building collapsed.

The Army Medical Command examined the recovered remains to determine the causes of death. Approximately one-third of the fatalities were related to the impact, one-third to the fire, and one-third to a combination of impact and fire.

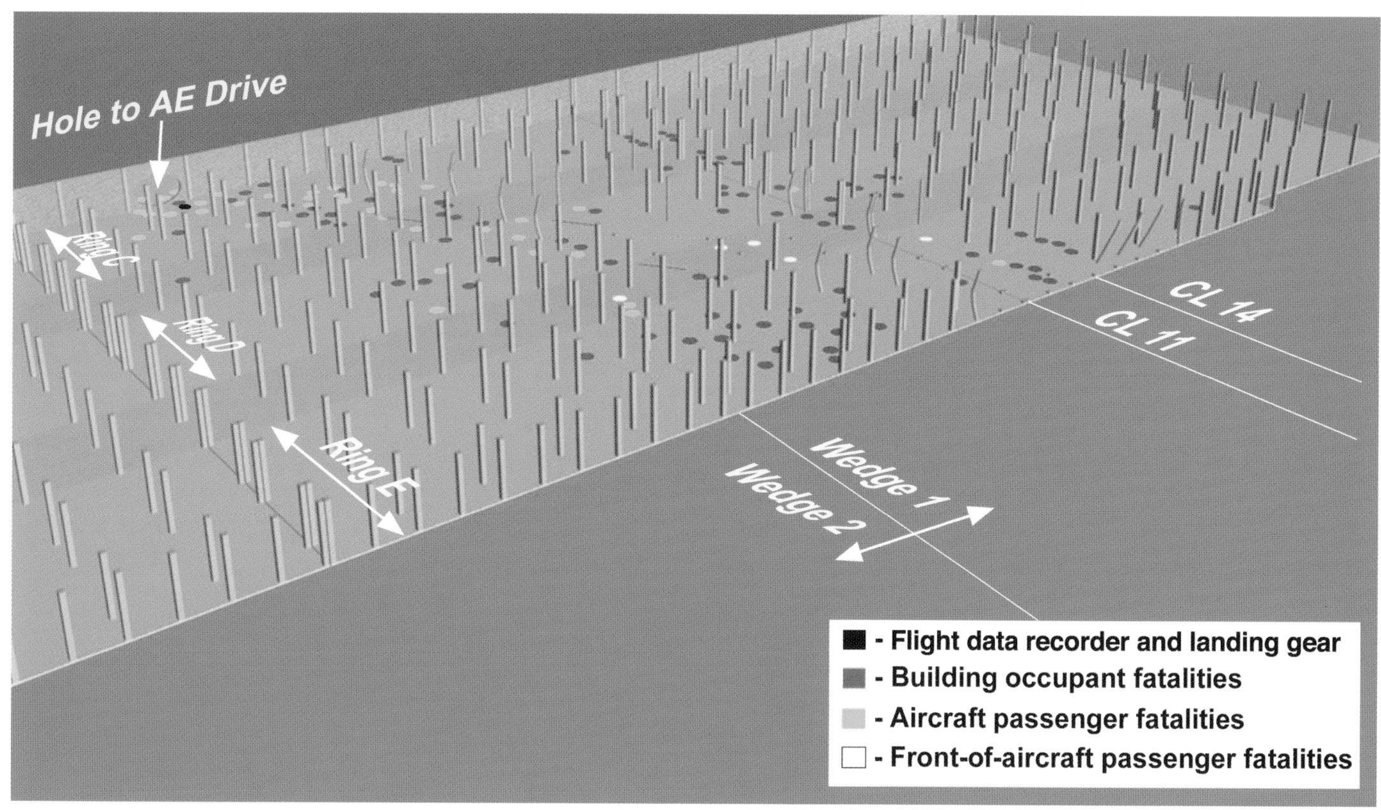

Figure 3.17 Fatalities found in first story

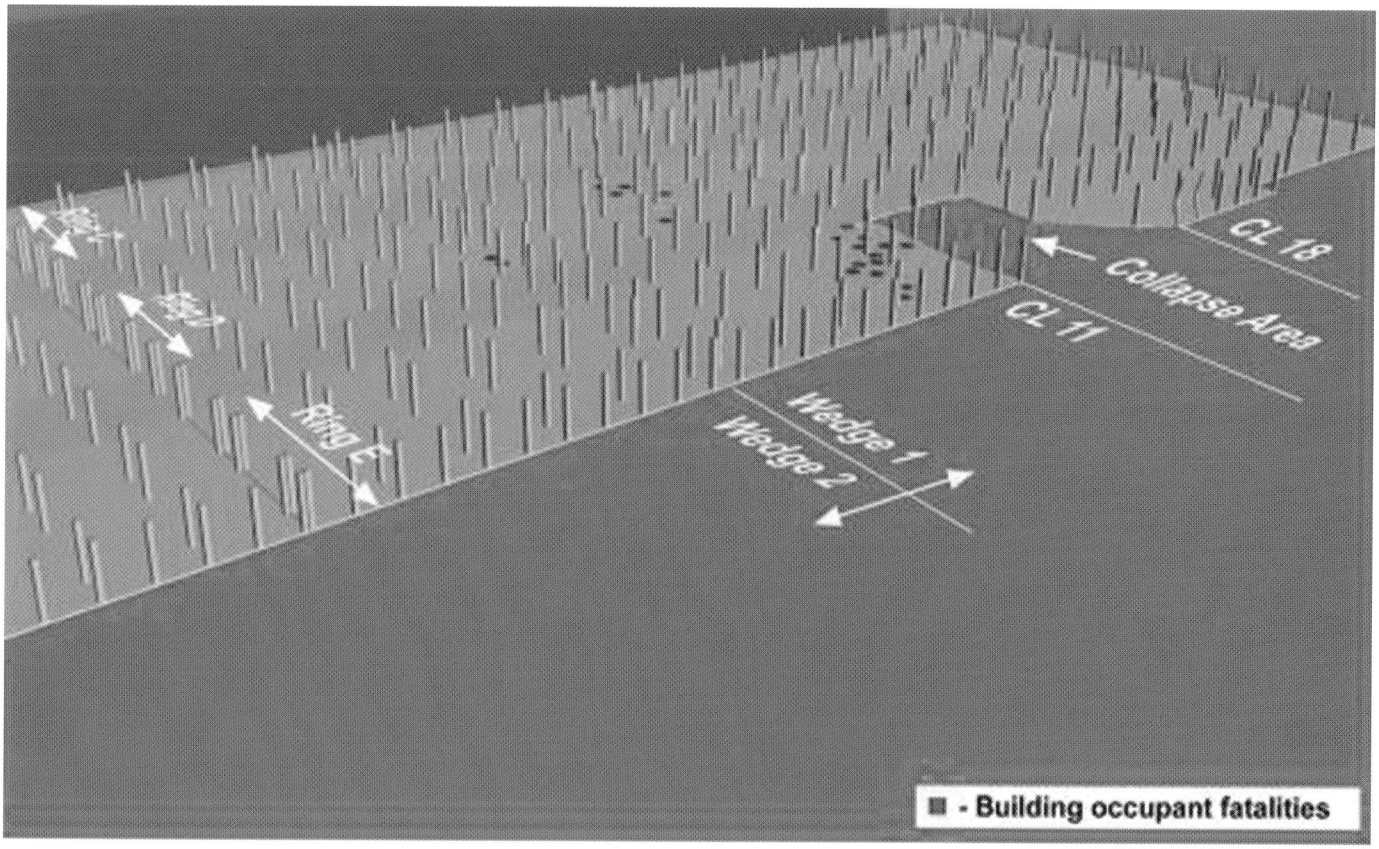

Figure 3.18 Fatalities found in second story

4. REVIEW OF OTHER STUDIES

Studies of the Pentagon crash have been or are being conducted by entities other than the BPS team. A number of these entities have shared information relevant to the goal of the building performance study. This information is described in 4.1–4.4.

4.1 RENOVATION ENGINEER

Following the events of September 11 the Pentagon Renovation Program initiated a building performance evaluation under the direction of the chief engineer, Georgine K. Glatz. The objective is, through building assessments and interviews with survivors, to learn how to maximize survival rates in similar events. The evaluation addresses fire suppression and rescue activities; building operations; human factors; and structural, architectural, fire-protection, mechanical, electrical, information management, and telecommunications systems. Dr. Glatz met with the BPS team on January 9, 2002, and with the team leader on a number of other occasions. The work of her task force to a great extent underlies the structural analyses set forth here and in section 7 and helps to explain the structural system within the context of the Pentagon's unique form and function.

4.2 RECONSTRUCTION ENGINEER

Immediately following the September 11 crash the Pentagon Renovation Program Office contracted with K.C.E. Structural Engineers, P.C., to oversee the recovery, demolition, and reconstruction of the site. As part of this effort, the principal, Allyn E. Kilsheimer, P.E., ordered materials tests on the concrete and the reinforcing steel. Two sets of cores were extracted for compressive stresses. In addition, numerous petrographic samples were taken and examined. Concrete testing and petrographic analyses were performed by ECS, Ltd., of Chantilly, Virginia, and reinforcing steel was tested by Arctech Testing, L.L.C., also of Chantilly, Virginia. The materials tests and examination data described below were provided to the BPS team for its use.

On October 2, 2001, the first set of six concrete cores was taken from beams along the west exterior wall in the debris of the collapsed area. Cores had nominal diameters of 3 in. and length-to-diameter ratios ranging from 1.73 to 2.09. When tested dry in accordance with American Society for Testing and Materials Standard C-42-99, corrected compressive strengths ranged from 4,420 to 5,310 psi. The mean and standard deviation for the tested compressive strengths were respectively 5,070 and 369 psi.

On October 29, 2001, 10 additional 3 in. diameter concrete cores were taken from an area to the north of the impact area, where the fire was severe. The purpose for testing these cores was to assist in the decision about how much of the building could be restored. These samples had length-to-diameter ratios ranging from 1.27 to 2.01, and corrected compressive strengths in this set of tests ranged from 2,180 to 4,210 psi. The mean and standard deviation for the tested compressive strengths were respectively 3,550 and 672 psi.

The test results for the two sets of cores differed substantially, both in the magnitudes of the average strengths and in the variability of results within each set. It is possible that these variations resulted from differences in mixing practices used at the several concrete batch plants that were established on the site when the Pentagon was originally constructed. It is also possible that the

Table 4.1 Test Results for Reinforcing Steel

Sample	Bar Size	Yield Strength[a] (psi)	Tensile Strength (psi)	Percentage Elongation
1	4	69,200	114,500	25
2	4	52,800	95,400	26
3	7	43,100	78,600	34
4	8	46,800	78,000	31
5	3	50,200	82,700	30

[a]Yield strength determined for 0.5 percent extension under load.

strength variations are the result of differences in exposures to heat during the fire that followed the impact of the aircraft.

According to ECS, Ltd., samples for petrographic evaluation were taken from the northern end of the fire-damaged area. Many of the samples exhibited symptoms of fire damage during petrographic examinations. The overall concrete quality was classified as fair to good, with approximately 13/16 in. carbonation. ECS, Ltd., determined that the concrete was placed originally with high slump.

Arctech Testing tested five samples of reinforcing steel. The results of their tests are summarized in table 4.1.

According to Arctech Testing, sample 1 passed all tensile strength requirements in table 2 of American Society for Testing and Materials (ASTM) Standard A615 for grade 60; samples 2 and 5 passed all tensile strength requirements in table 2 of ASTM A615 for grade 40 (and would also for the former grade 50); and samples 3 and 4 met the yield and tensile strength requirements in table 2 of ASTM A615 for grade 40.

KCE informed the Pentagon BPS team that actual construction deviated from construction documents in many locations.

4.3 CORPS OF ENGINEERS STUDY

With the cooperation of the Army Corps of Engineers the team reviewed a study that focused on protecting Pentagon occupants and ensuring mission continuity in the wake of the September 11 crash.

The study noted that the windows demonstrated desirable energy absorption in their response to the blast and also withstood the impact of debris fragments. The steel framework supporting these windows responded with strength and ductility, as intended in its design.

The study further noted that the reinforced-concrete frame of the building is inherently robust. This stems from the closely spaced spiral reinforcement in the columns in the lower stories. Additionally, much of the beam and girder flexural reinforcement continued through the support connection. The robustness was demonstrated in the resistance to progressive collapse following the crash.

4.4 URBAN SEARCH AND RESCUE ENGINEERS

Approximately a dozen professional structural engineers participated on-site in the initial rescue and recovery operations between September 11 and September 21. While their focus was on emergency search and rescue, they observed the initial structural condition of the building. Their photographs and other documentation proved valuable in understanding the condition of the building.

5. BPS SITE INSPECTIONS

Members of the BPS team inspected the site on two occasions. Between September 14 and September 21, 2001, team leader Paul Mlakar had limited access to the site while rescue and recovery operations were still in progress. On this early inspection visit, he examined the exterior of the building and portions of the building interior.

Controlled access to the site was granted to the full team after rescue and recovery operations were complete. On October 4, 2001, the Pentagon team, together with John Durrant, the executive director of ASCE's institutes, and W. Gene Corley, the BPS team leader at the World Trade Center, inspected the interior and exterior of the damaged area of the Pentagon for approximately four hours.

The inspection of the BPS team focused on obvious physical damage, primarily in the region of the impact. This inspection was not comprehensive. It did not address fire damage to concrete as a material, and it did not result in full documentation of all physical damage or as-built construction.

By the time the full Pentagon BPS team visited the site, all debris from the aircraft and structural collapse had been removed (figure 5.1) and shoring was in place wherever there was severe structural damage. The design team charged with reconstructing the Pentagon was assessing the building and preparations were being made to demolish the areas for reconstruction. Consequently, the Pentagon BPS team never had direct access to the structural debris as it existed immediately after the aircraft impact and subsequent fire.

Following a brief period of orientation, the Pentagon BPS team members paired off into four separate inspection teams to survey different portions of the damaged area of the building. Three inspection teams documented conditions in the first story, and the fourth team documented the upper stories.

For the purposes of the inspection, damage to individual structural elements was classified as follows:

- No significant damage;
- Cracking and spalling, but no significant impairment in function;
- Heavy cracking and spalling, with some impairment in function (the member remaining straight or nearly so);
- Large deformation, with significant impairment in function;
- Members missing, broken, disconnected, or otherwise without remaining function.

The teams attempted to inspect and photograph all columns with significant visible damage and most of the beams and floor bays with significant visible damage. To the extent possible, it was noted whether physical loads or the effects of fire caused the observed damage. The BPS team also noted the performance of windows and exterior wall reinforcements that had been installed to enhance blast resistance in Wedge 1 prior to the attack. However the BPS team inspections were not comprehensive, and they did not address fire-related material degradation.

The collapsed portion of Ring E was immediately south of an expansion joint on column line 11 (figures 5.2 and 5.3). The collapsed area extended south from the expansion joint to

Figure 5.1 Impact area with debris removed

Figure 5.2 Beams and columns on north side of expansion joint

Figure 5.3 Aerial view of collapsed portion of Ring E

Figure 5.4 Remaining structure at southern extreme of collapse area

Figure 5.5 Missing first-floor columns 11A, 11B, and 11C

Figure 5.6 Facade damage to the north of the impact area

Figure 5.7 First-floor facade damage to the north of the impact area

approximately column line 15 on the east side of Ring E and to approximately column line 18 on the west side of Ring E (figure 5.4). No portion of Ring D or Ring C collapsed; nor did either of the two-story sections between the rings.

Since all debris was removed prior to the detailed inspection, the team was unable to determine specifically the level and extent of impact damage in this region of the building.

In general, the first-floor interior columns were severely damaged immediately adjacent to the collapse area on the north side of the expansion joint on column line 11 in Ring E. First-floor columns 11A, 11B, and 11C to the north of the expansion joint were missing (figure 5.5). Upper columns on the north side of the expansion joint on column line 11 were intact, except for the second-floor columns at 11A and 11B. These columns were severed at the second floor, which was also damaged at this location.

None of the facade in the collapse area was accessible for inspection. However, the team did observe that limestone of the first-floor facade was seriously damaged to the north to column line 8 (figure 5.6). Some first-floor limestone panels of the facade were missing for an additional 30 to 50 ft to the north (figure 5.7).

The first-floor exterior column on column line 9 remained in place, but the rest of the exterior columns south to column line 11, at the start of the collapsed area, were gone.

To the south, facade panels on both the first and second floors between column lines 18 and 20 were severely damaged (figure 5.8).

The exterior of the building showed clear evidence of the extensive fire that occurred within the building. The limestone facade was blackened by smoke for more than 200 ft to the north of the impact point (figure 5.9). Evidence of fire damage was less severe to the south (figure 5.10), and even immediately adjacent to the impact area the facade to the south showed little evidence of fire damage.

The west facade of the Pentagon was severely scarred by debris impact, particularly to the south of the collapse area (figure 5.11). Just above the second-floor

Figure 5.8 Facade damage to the south of the impact area

Figure 5.9 Exterior evidence of fire to the north of the impact area

Figure 5.10 Exterior evidence of fire to the south of the impact area

Figure 5.11 Debris scars at upper levels south of impact area

slab, the exterior columns on column lines 18 and 19 exhibited aligning gashes that seem to indicate impact by the right wing of the aircraft (figure 5.12). An area of broken limestone of the facade over the exterior column on column line 20 also aligned with these gashes. The fire station to the north of the heliport and the impact area was also damaged by flying debris (figure 5.13).

The team observed that the upgraded window system was generally still in place within the reinforced frames (figure 5.14). Windows that had not been upgraded generally were broken for several hundred feet to the north of the impact point (figure 5.15).

The aircraft had entered the building at an angle, traveling in a northeasterly direction. With the possible exception of the immediate vicinity of the fuselage's entry point at column line 14, essentially all interior impact damage was inflicted in the first story: The aircraft seems for the most part to have slipped between the first-floor slab on grade and the second floor. The path of damage extended from the west exterior wall of the building in a northeasterly direction completely through Ring E, Ring D, Ring C, and their connecting lower floors. There was a hole in the east wall of Ring C, emerging into AE Drive, between column lines 5 and 7 in Wedge 2 (figure 5.16). The wall failure was approximately 310 ft from where the fuselage of the aircraft entered the west wall of the building. The path of the aircraft debris passed approximately 225 ft diagonally through Wedge 1 and approximately 85 ft diagonally through a portion of Ring C in Wedge 2.

Columns and beams along the path of the debris and within the fire area were damaged to varying degrees. Some columns and beams were missing entirely (figure 5.17), while others nearby sometimes appeared unscathed.

Most of the serious structural damage was within a swath that was approximately 75 to 80 ft wide and extended approximately 230 ft into the first floor of the building. This swath was oriented at approximately 35 to 40 degrees to the perpendicular to the exterior wall of the Pentagon. Within the swath of serious damage was a narrower, tapering area that contained most of the very severe structural damage. This tapering area approximated a triangle in plan and had a width of approximately 90 ft at the aircraft's entry point and a length of approximately 230 ft along the trajectory of the aircraft through the building.

Figure 5.12 Gashes from impact of right wing

Severe damage included heavy cracking and spalling, either from impact or from the ensuing fire. The concrete cover had been completely dislodged from the spirally reinforced core concrete and steel of the most heavily damaged columns that remained in place. Figure 5.18 shows a column at the edge of the collapse area that appears to have been stripped of most of its cover by impact. Figure 5.19 shows a column with impact damage and relatively severe fire damage.

Several columns were substantially distorted, exhibiting lateral displacement at the column midheight equal to at least three times the diameter of the spiral cage. Some highly distorted columns were bent in uniform curvature with discrete hinges at each end (figure 5.20), while others were bent into triple curvature (figure 5.21). In these cases, the vertical column steel remained attached to the foundation below and the second-floor beams above (figure 5.22). The deformed shapes of the columns with this damage were smooth curves: generally they did not have discrete deformation cusps.

Figure 5.13 Damage to fire station

Figure 5.14 Upgraded windows in place

Figure 5.15 Broken conventional windows

Figure 5.16 Hole to AE Drive

Federal Bureau of Investigation

Figure 5.17 Column missing at location 3K

Figure 5.18 Column 15B with concrete cover removed by impact

Figure 5.19 Column 3N damaged by impact and fire

Figure 5.20 Column 3L with large deformation and discrete hinges

Figure 5.21 Column with triple curvature

Figure 5.22 Top of column 3L, still attached to underside of second-floor framing

Figure 5.23 Severed column 5H

Figure 5.24 Cracked column 5N

Figure 5.25 Impact and fire damage near column line C

Figure 5.26 Impact damage near column 6K

Figure 5.27 Breached second-floor slab

Figure 5.28 Crack patterns in second-floor slab

Figure 5.29 Concrete stripped from floor beam

Figure 5.30 Missing beam in air shaft

Figure 5.31 Fire damage to column in second story

In the worst cases, first-floor columns were severed from the second floor above or from the slab on grade or were missing entirely. Severed columns generally were lying on the slab on grade, still attached to the floor (figure 5.23). These columns were straight (except for the discrete bends at the connections to the floor) in their prone positions.

The orientations of the distorted columns and the columns that were severed all indicated a common direction for the loads that caused the damage. The direction of column distortion consistently formed an angle of approximately 42 degrees with the normal to the west exterior wall of the Pentagon.

Most first-floor columns outside of the direct path of the moving debris had no visible damage or had light cracking and spalling (figure 5.24), most probably caused by fire.

There were two primary areas where beams and slabs of the second floor were damaged by impact: (1) in an area bounded approximately by column lines A, 5, B, and 11 (figure 5.25) and (2) in an area bounded approximately by column lines E, 5, H, and 7 (figure 5.26). The most severe damage was in the second of these areas, where the second-floor slab was breached and pushed upward approximately 18 in. (figure 5.27). In other panels in these regions, slab cracks were relatively closely spaced (6 to 8 in. apart) in ring and radial patterns (figure 5.28). Some floor beams were completely stripped from the underside of the slab above (figure 5.29). In addition, second-floor beams framing air shafts to the low roof between Ring C and Ring D were severely damaged (figure 5.30).

Damage to the structure above the second floor (outside the collapse area) appeared to be related to fire (figure 5.31) rather than impact, even though the orientation of the region of severe damage aligned generally with severe damage below the second-floor slab. Fire damage in the second story appeared most severe around the region of collapse and near the breach in the second-floor slab.

Generally, the most obvious fire damage was between the fire walls to the north and south of the area directly damaged by the aircraft debris. The most severe fire damage occurred on the first and second floors. The team noted no impact damage above the second story.

Appendix B includes a summary of damage and a photograph of each first-story column in the area of impact damage.

6. DISCUSSION

6.1 IMPACT DAMAGE

The site data indicate that the aircraft fuselage impacted the building at column line 14 at an angle of approximately 42 degrees to the normal to the face of the building, at or slightly below the second-story slab. Eyewitness accounts and photographs taken by a security camera suggest that the aircraft was flying on nearly a level path essentially at grade level for several hundred feet immediately prior to impact. Gashes in the facade above the second-floor slab between column lines 18 and 20 to the south of the collapse area suggest that the aircraft had rolled slightly to the left as it entered the building. The right wing was below the second-floor slab at the fuselage but above the second-floor slab at the tip, and the left wing struck the building entirely below the second-floor slab, to the north of column line 14.

The width of the severe damage to the west facade of the Pentagon was approximately 120 ft (from column lines 8 to 20). The projected width, perpendicular to the path of the aircraft, was approximately 90 ft, which is substantially less than the 125 ft wingspan of the aircraft (figure 6.1). An examination of the area encompassed by extending the line of travel of the aircraft to the face of the building shows that there are no discrete marks on the building corresponding to the positions of the outer third of the right wing. The size and position of the actual opening in the facade of the building (from column line 8 to column line 18) indicate that no portion of the outer two-thirds of the right wing and no portion of the outer one-third of the left wing actually entered the building.

It is possible that less of the right wing than the left wing entered the building because the right wing struck the facade crossing the level of the second-floor slab. The strength of the second-floor slab in its own plane would have severed the right wing approximately at the location of the right engine. The left wing did not encounter a slab, so it penetrated more easily.

In any event, the evidence suggests that the tips of both wings did not make direct contact with the facade of the building and that portions of the wings might have been separated from the

Figure 6.1 Aircraft aligned with damage on west facade

fuselage before the aircraft struck the building. This is consistent with eyewitness statements that the right wing struck a large generator before the aircraft struck the building and that the left engine struck a ground-level, external vent structure. It is possible that these impacts, which occurred not more than 100 ft before the nose of the aircraft struck the building, may have damaged the wings and caused debris to strike the Pentagon facade and the heliport control building.

The wing fuel tanks are located primarily within the inner half of the wings. The center of gravity of these tanks is approximately one-third of the wing length from the fuselage. Considering this tank position and the physical evidence of the length of each wing that could not have entered the building, it appears likely that not more than half of the fuel in the right wing could have entered the building. While the full volume of the left wing tank was within the portion of the wing that might have entered the building, some of the fuel from all tanks rebounded upon impact and contributed to the fireball. Only a portion of the fuel from the left and right wing tanks and the center fuselage tank actually entered the building.

The height of the damage to the facade of the building was much less than the height of the aircraft's tail. At approximately 45 ft, the tail height was nearly as tall as the first four floors of the building. Obvious visible damage extended only over the lowest two floors, to approximately 25 ft above grade.

Damage to the first-floor columns is summarized in figures 6.2 and 6.3. In formulating opinions about columns in the collapse area, the BPS team interpreted photographs taken after impact and before collapse.

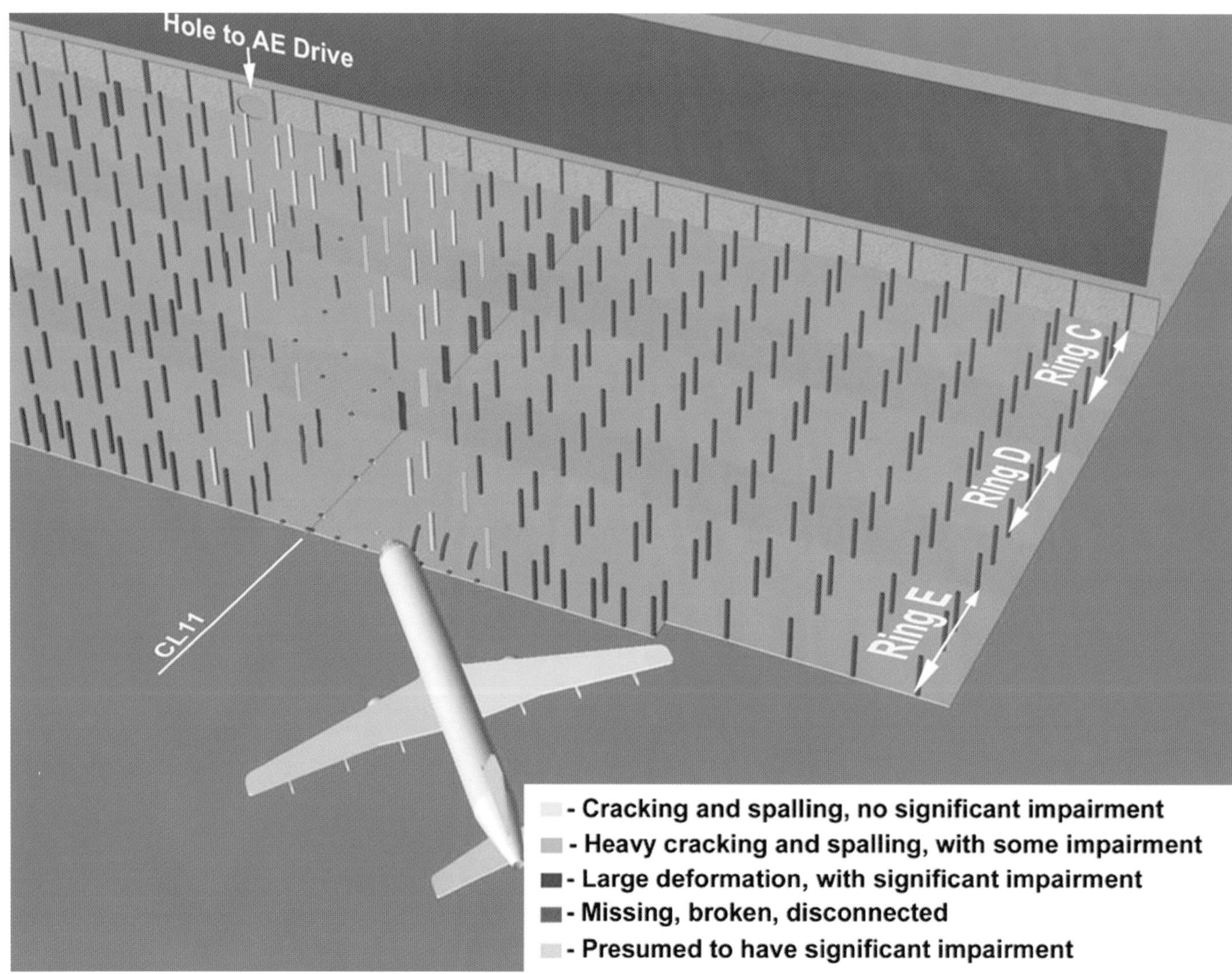

Figure 6.2 Damage to columns in first story viewed along path of aircraft

The team members do not have direct information on the impact damage to the upper floors in the collapsed portion of the building. However, based on observations of the condition of the adjoining structure and the photographs of the building before the collapse, the following general observations may be made:

Impact damage on the first floor was extensive near the entry point of the aircraft. It is likely that the exterior first-floor columns from column line 10 to column line 14 were removed entirely by the impact and that the exterior columns on column lines 9, 15, 16, and 17 were severely damaged. Most probably, many or most of the first-floor interior columns in the collapse area were heavily damaged by impact.

The removal of the second-floor exterior column on column line 14, probably by the fuselage tail, suggests that the second-floor slab in this area was also severely damaged even before the building collapsed. In the portion of the building that remained standing to the north of the expansion joint, the slab and second-floor columns at column lines A, B, and C were heavily damaged. This condition, which is consistent with the trajectory of the aircraft, suggests that the second-floor slab from the expansion joint on column line 11 south to the fuselage entry point on column line 14—including columns 11B, 11C, and 13A on the second floor—was heavily damaged, perhaps destroyed.

It is difficult to judge the condition of other columns on the second floor in the collapse area. However, more likely than not column 15A was relatively undamaged. It is unlikely that columns above the second floor sustained impact damage, even in the area that ultimately collapsed.

Figure 6.4 summarizes the damage to the second-floor beams. (Lightly damaged beams were numerous, but they are not shown in this figure because the BPS team was not able to complete a comprehensive survey during its inspections.) Figure 6.5 summarizes breaches in the second-floor slab and damage to columns that supported the third floor.

Impact damage to the structure above the second-floor slab did not extend more than approximately 50 ft into the building. This shows that the aircraft slid between the first-floor slab on grade and the second-floor slab for most of its distance of travel after striking the building.

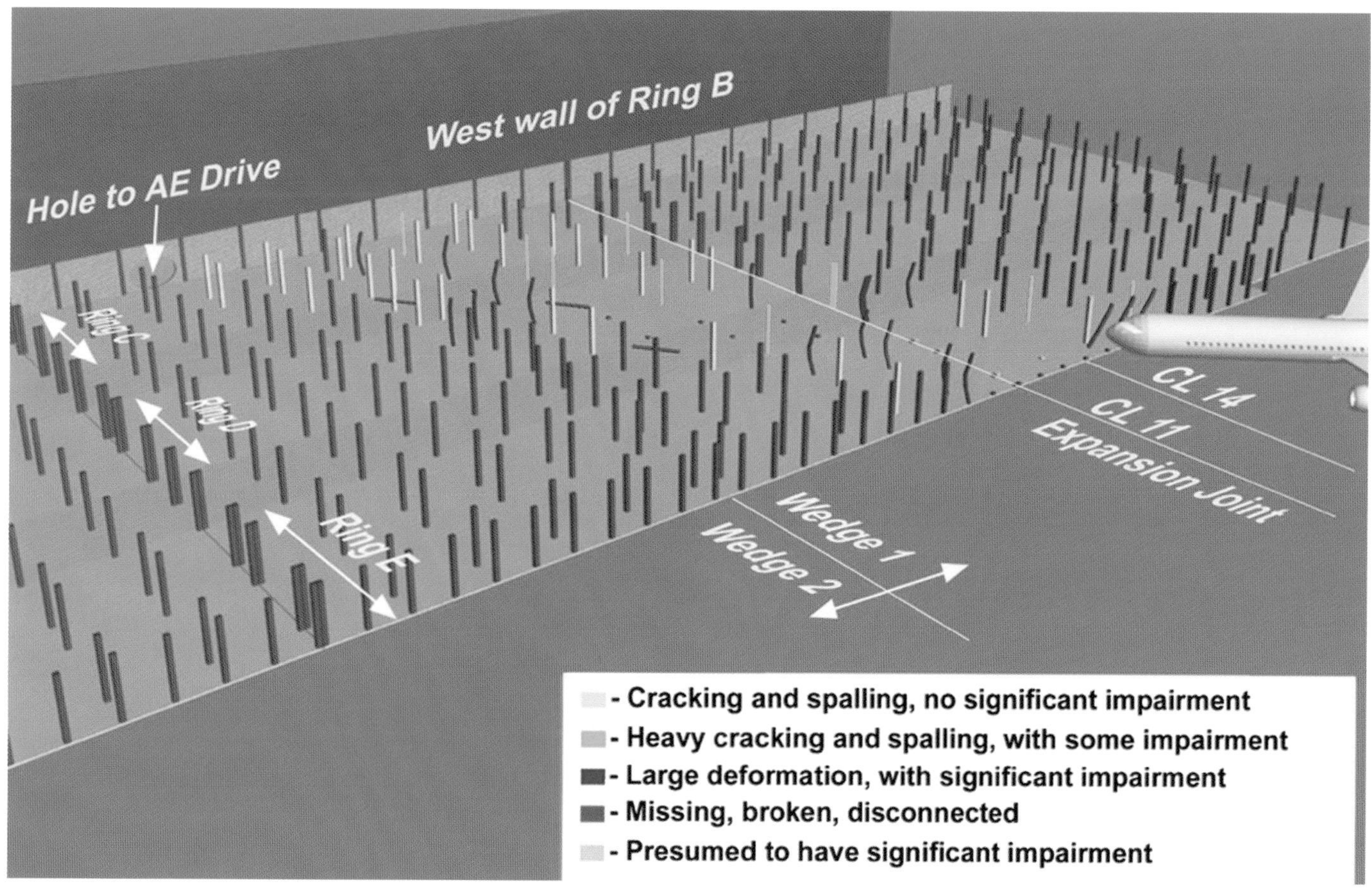

Figure 6.3 Damage to columns in first story viewed across path of aircraft

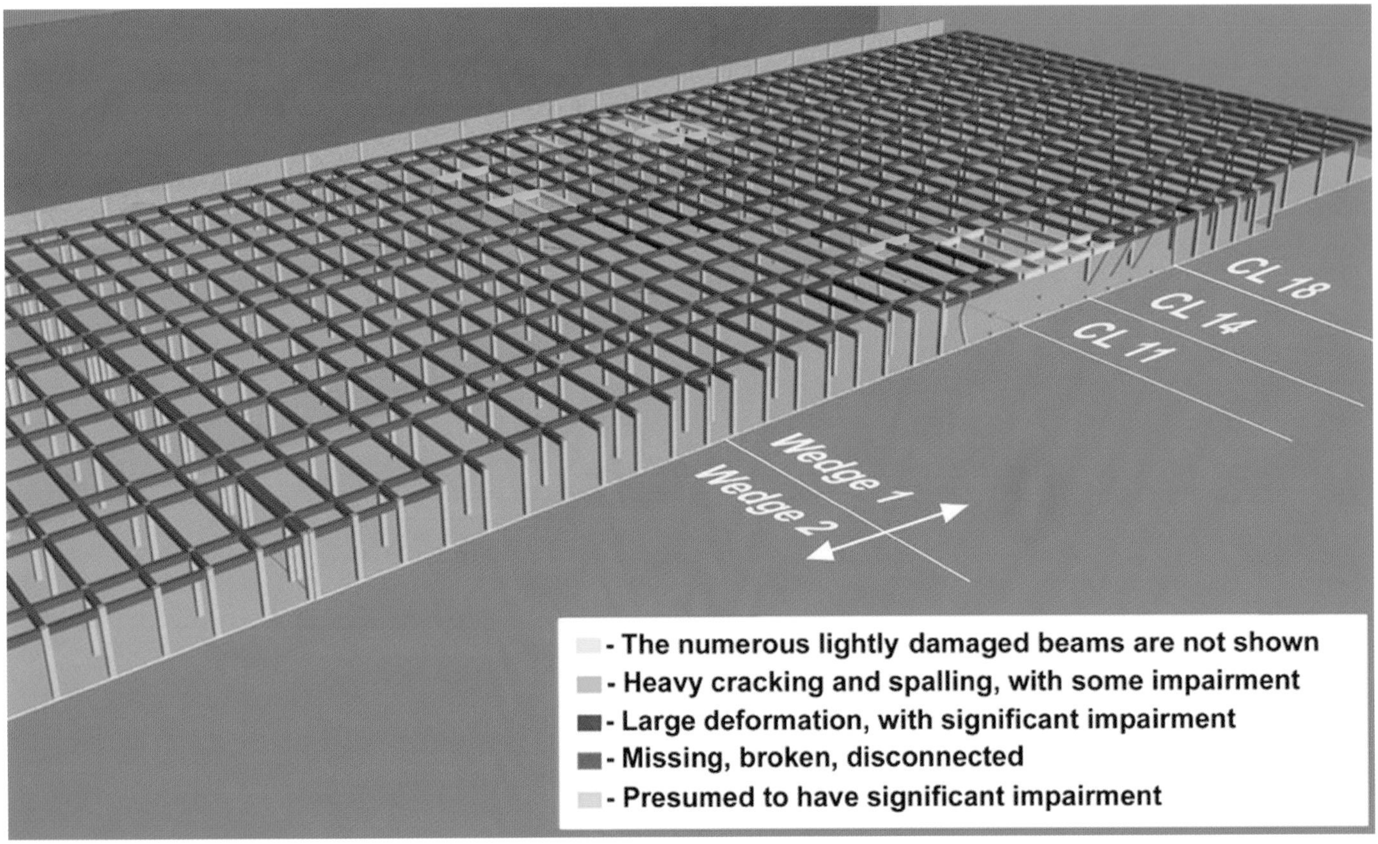

Figure 6.4 Damage to second-story beams

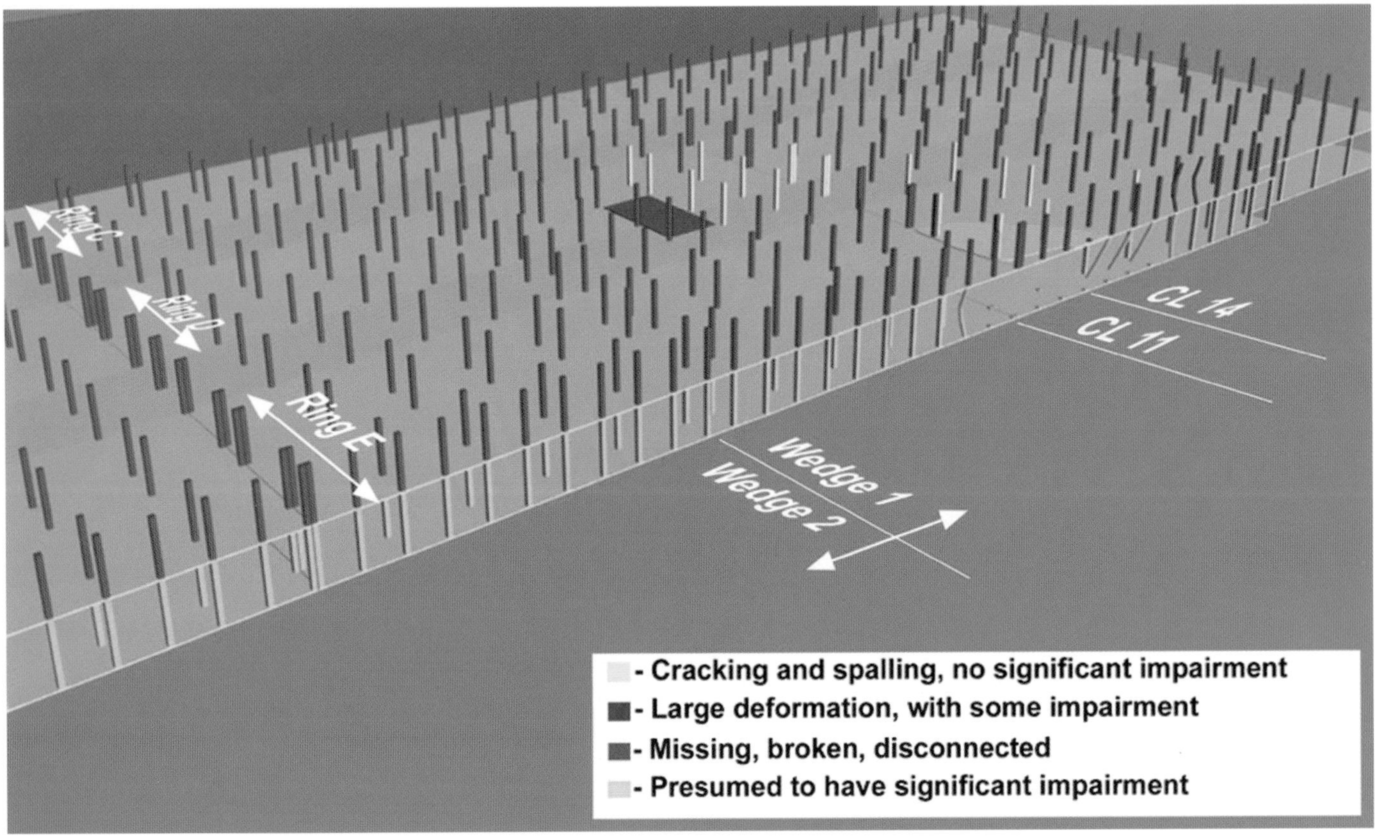

Figure 6.5 Damage to columns and slab in second story

Along the path of the movement of aircraft debris through the building, the most severe damage was confined to a region that can be represented approximately by a triangle centered on the trajectory of the aircraft in plan, with a base width at the aircraft entry point of approximately 90 ft and a length along the aircraft path of approximately 230 ft (figure 6.6). However, within this triangular damage area there were a few relatively lightly damaged columns interspersed with heavily damaged columns along the path of the aircraft debris through the building. Column 1K (figure 6.6), located 200 ft from the impact point, was the last severed column along the path of the aircraft. Note that columns on grids E and K are much weaker than the other columns because they support only one floor and a roof.

There were two areas of severe impact damage in the first story. The first area along the path of the aircraft was within approximately 60 ft of the impact point and corresponds generally to the area that collapsed. In the collapse area and for approximately 20 ft beyond the collapse area along its northern and eastern edges, columns were removed or very severely damaged by impact. In addition, there was serious second-floor beam and slab damage for 60 ft to the north of the collapse area, especially along a strip bounded approximately by column lines B and C.

The second area of severe damage was bounded approximately by column lines E, 5, G, and 9. In this region, which was beyond a field of columns that remained standing, several columns were severed and there was significant second-floor beam and slab damage. In both areas, severe slab damage appeared to be caused by moving debris rather than by overpressure from a blast.

In an effort to characterize the influence of the aircraft on the structure and, by extension, to characterize the loads on the structure, the team analyzed the available data to extract information about the destruction of the aircraft.

Most likely, the wings of the aircraft were severed as the aircraft penetrated the facade of the building. Even if portions of the wings remained intact after passing through the plane of the facade, the structural damage pattern indicates that the wings were severed before the aircraft penetrated more than a few dozen feet into the building. Ultimately, the path of the fuselage debris passed between columns 9C and 11D (figure 6.6), which were separated by approximately 28 ft at a depth of approximately

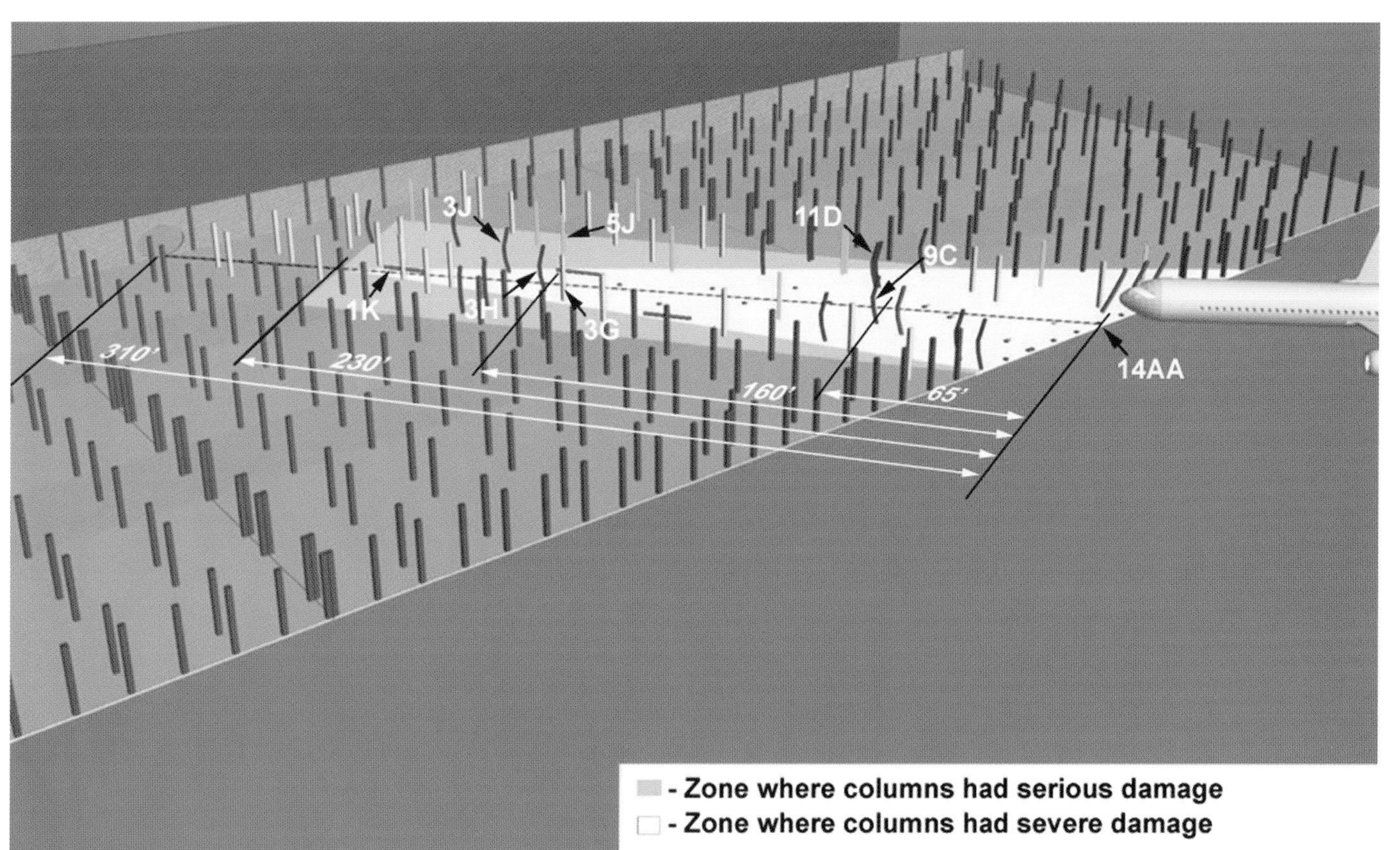

Figure 6.6 Damage regions in first story

Test Columns after Three-Hour Exposure and Time-Temperature Curves for ASTM E-119 and ISO 834

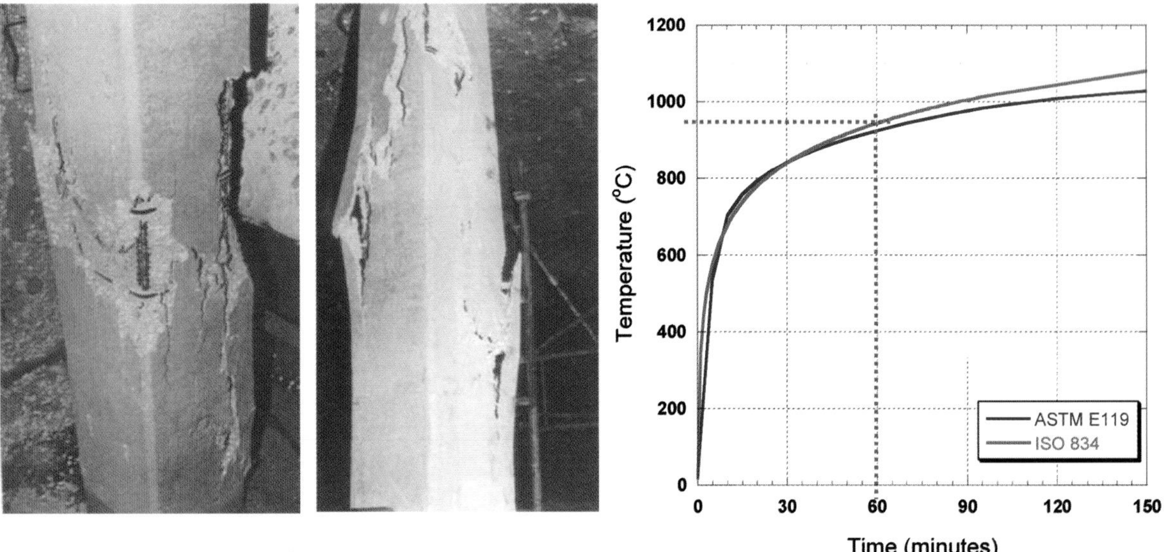

Figure 6.7 Test columns after three-hour exposure

65 ft along the aircraft's path. Columns 9C and 11D were severely distorted but still in place: hence the wings clearly did not survive beyond this point.

At a depth of approximately 160 ft into the building, columns 3G, 3H, 3J, and 5J (figure 6.6) were damaged but still standing, although in the direct path of the fuselage. With a maximum spacing of less than 14 ft between pairs of these columns in a projection perpendicular to the path of the fuselage, it is highly unlikely that any significant portion of the fuselage could have retained structural integrity at this point in its travel. More likely, the fuselage was destroyed much earlier in its movement through the building. Therefore, the aircraft frame most certainly was destroyed before it had traveled a distance that approximately equaled the length of the aircraft.

The debris that traveled the farthest traveled approximately twice the length of the aircraft after entering the building. To come to rest at a point 310 ft (figure 6.6) from the area of impact at a speed of 780 ft/s, that debris experienced an average deceleration of approximately 30*g*.

The influence of the structure on the deceleration of the aircraft (and, conversely, the influence of the aircraft on the structure) can be appreciated by comparisons with examples of aircraft belly-landed in controlled circumstances. In 1984, the Federal Aviation Administration (FAA) conducted a controlled impact demonstration (Department of Transportation 1987) to evaluate the burn potential of antimisting kerosene fuel. In that test, the FAA landed a Boeing 720 aircraft (weighing approximately 175,000 lb) without landing gear on a gravel runway at Edwards Air Force Base. The aircraft in that test was flying at approximately 250 ft/s when it made first contact, but it slid approximately 1,200 ft before it stopped. Although the test aircraft was traveling at approximately one-third the speed of the aircraft that struck the Pentagon, its sliding distance was approximately 3.9 times that of the Pentagon attack aircraft. Clearly, the short stopping distance for the aircraft striking the Pentagon derived from the energy dissipated through the destruction of the aircraft and building components; the acceleration of building contents; the loss of lift when the wings were severed from the aircraft; and effective frictional and impact forces on the first-floor slab, the underside of the second-floor slab, and interior columns and walls.

A study of the locations of fatalities also yields insight into the breakup of the aircraft and, therefore, its influence on the structure. The remains of most of the passengers on the aircraft were found near the end of the travel of the aircraft debris. The front landing gear (a relatively solid and heavy object) and the flight data recorder (which had been located near the rear of the aircraft) were also found nearly 300 ft into the structure. By contrast, the remains of a few individuals (the hijacking suspects), who most likely were near the front of the aircraft, were found relatively close to the aircraft's point of impact with the building. These data suggest that the front of the aircraft disintegrated essentially upon impact but, in the process, opened up a hole allowing the trailing portions of the fuselage to pass into the building.

Several columns exhibited severe bends. However, the predominant evidence suggests that these columns generally did not receive impact from a single, rigid object. Instead, the deformed shapes of these columns are more consistent with loads that were distributed over the height of the columns.

The analyses of the available data reveal that the wings severed exterior columns but were not strong enough to cut through the second-floor slab upon impact. (The right wing did not enter the building at the point where it struck the second-floor slab in its plane.) The damage pattern throughout the building and the locations of fatalities and aircraft components, together with the deformation of columns, suggest that the entire aircraft disintegrated rapidly as it moved through the forest of columns on the first floor. As the moving debris from the aircraft pushed the contents and demolished exterior wall of the building forward, the debris from the aircraft and building most likely resembled a rapidly moving avalanche through the first floor of the building.

6.2 FIRE DAMAGE

Fire damage generally was similar to that normally resulting from serious fires in office buildings. Clearly, some of the fuel on the aircraft at impact did not enter the building, either because it was in those portions of the wings that were severed by the impact with the facade or with objects just outside of the building, or because it was deflected away from the building upon impact with the facade; that fuel burned outside the building in the initial fireball. Generally, fire damage to columns, beams, and slabs was limited to cracking and spalling in the vicinity of the aircraft debris. There were two areas with more severe damage. One area, to the north of the path of the aircraft, was bounded approximately by column lines 4, 7, A, and D. The other area, to the south of the path of the aircraft, was in the vicinity of column lines K and L and crossing column lines 11, 12, and 13. In both areas,

Figure 6.8 First-story columns 5N and 5M

Figure 6.9 First-story columns 3M and 3N

there was more serious spalling and cracking than occurred typically throughout the fire area. Fire damage on the second floor in the vicinity of the path of the aircraft was generally more severe than in the same areas directly below on the first floor.

Comparing the state of damage in these columns with damage observed in laboratory fire tests of reinforced-concrete columns—made with concrete of similar compressive strength and subjected to well-defined heating regimes—may provide an indication of the lower bound of the maximum temperature reached at different locations in the Pentagon.

Figure 6.7 shows two 11.5 by 11.5 in. test columns that failed after more than three hours of exposure to standard fire ISO 834 (similar to ASTM E-119). The tests were conducted in Belgium at the University of Liège and reported in Phan et al. (1997). The columns were made of C20 concrete (2,900 psi [20 MPa]), which is within the range of compressive strengths measured for concrete at the Pentagon. They differed in the reinforcements: one had eight 1/2 in. diameter longitudinal rebars and the other had four 1 in. diameter longitudinal rebars. Both had a concrete cover of 1.2 in. Both columns carried an axial load of 50 percent of the room temperature capacity prior to fire exposure, and the loads were maintained throughout the heating process. The authors observed that longitudinal cracks and corner spalling, which are typical in concrete columns because of the large thermally induced transverse strains that are unrestrained in the transverse direction, occurred approximately one hour after exposure to ISO 834. This coincides with an ambient temperature of about 1,740°F (950°C).

Several structural elements in bays adjacent to the path of aircraft impact in the first floor did not sustain damage by impact. Rather, the damage to these elements was due to fire exposure. Because the structural elements in the Pentagon are believed to have had additional fire protection provided by the interior finishes—while the laboratory columns were fully exposed—and because the rate of temperature rise in the actual fire is believed to be greater than that prescribed by ISO 834, the comparison is not exact. However, it should provide an indication of the lower bound of the temperature at some locations in the Pentagon.

Figures 6.8 and 6.9 show first-floor columns 5M, 5N, 3M, and 3N at the time of the BPS team visit. These columns are located in Ring C, toward the end of the damage path. All four sustained

Figure 6.10 First-story columns 7J, 7K, and 7L

thermal damage in the form of longitudinal cracks and corner spalling. Some sections of the columns appeared blackened, probably as a result of direct exposure to flame caused by partial loss of interior finishes. It took a little more than one hour of exposure to ISO 834—at a corresponding ambient temperature of about 1,740°F (950°C)—for the longitudinal cracks and corner spalling to develop in the laboratory test columns. This indicates that the temperature of the fire at this location might have reached a similar level.

Similar damage was also observed for columns 7J, 7K, and 7L—seen in figures 6.10 and 6.11—suggesting that the maximum temperature at these locations might also have reached 1,740°F (950°C).

Fire damage to the underside of the second-floor slab at some locations can also be compared with laboratory tests for an indication of the lower bound of the maximum temperature reached. Figures 6.12, 6.13, and 6.14 show the state of the fire damage, which included extensive thermal cracks and edge spalling of supporting beams and girders on the underside of second- and third-floor slabs. In laboratory fire tests of concrete slabs conducted by Construction Technology Laboratories, Inc. (Shirley, Burg, and Fiorato 1987), which exposed the undersides of four high-strength concrete slabs and one normal-strength concrete slab to an ASTM E-119 standard fire for four hours, it was reported that the normal-strength concrete slab attained a fire-endurance rating of 88 minutes, based on the temperature rise on the unexposed surface criterion. Eighty-eight minutes of exposure to ASTM E-119 is equivalent to a maximum ambient temperature of about 1,832°F (1,000°C). The exposed and unexposed surfaces of the test slab were monitored periodically throughout the four-hour test fire. The authors reported that no cracks were visible on the exposed surface of any slabs during the fire test. After cooling, the specimens did exhibit reflective cracking, which included hairline cracks perpendicular to the slab edge at random locations along specimen perimeters. None of the slabs exhibited any spalling of the concrete surface during the test. As shown in figures 6.12–6.14, thermal cracks and edge spalling occurred on the undersides of some sections of the second- and third-floor slabs in the Pentagon. This indicates a more severe fire exposure at these locations; thus the standard fire exposure ASTM E-119 may prescribe the lower bound of the time-temperature curve for the real fire at these locations.

Figure 6.11 First-story column 7J

Figure 6.12 Fire damage to third-floor (light well) beams

6.3 EXTERIOR WALL UPGRADES

The structural upgrades of the exterior wall performed reasonably well, considering that they were not specifically designed for aircraft impact. The only window frames removed by the impact were those struck directly by the wings or the fuselage. On the second floor, immediately adjacent to where the fuselage entered the building, upgraded windows remained in their frames even though the surrounding masonry facade was completely removed.

Upgraded glass was generally not broken immediately after the impact or after the ensuing fire had been extinguished. By contrast, most of the original windows in a vast area of Wedge 2 were broken after the fire was extinguished. It is probable that some of these windows were broken by the fire or by firefighting efforts rather than by the effects of the impact.

Figure 6.13 Concrete stripped from second-floor beam at grid 9 between grids A and B; extensive cracking in slab

Figure 6.14 Fire-damaged second-floor beams

7. ANALYSIS

The work detailed in the previous section suggests that there are three issues of structural performance that require analysis. First, the impact of the aircraft laterally loaded a large number of spirally reinforced columns. It is important to note that the response ranged from complete removal of the columns to inconsequential damage. Second, a portion of the structure in which many of the columns had been destroyed by the impact remained standing. As such performance is to be desired, the reasons for it are of interest to the engineering profession. Third, a limited collapse occurred roughly 20 minutes after the impact of the plane. This calls for an examination of the fire loading of the structure in this interval. Comprehensive analyses of these three phenomena could be performed, but the fidelity of the available input and response data is not on a par with the demands of such attempts. The following sections contain quantitative data based on simple calculations to provide a perspective on the toughness of the structure and the effect of the fire.

7.1 RESPONSE OF COLUMNS TO IMPACT

The structural elements of the Pentagon that bore the brunt of the airplane impact were the first-story columns. The locations of the columns in the area affected by the impact and the ensuing fire are shown in figures 6.2 and 6.3 (section 6). All columns in the first story had square cross sections and spirally reinforced cores with a concrete cover of $1\,^{1}/_{2}$ in. The story height was 14 ft 1 in. There were two different arrangements of longitudinal reinforcement as shown in figure 7.1. The side dimensions varied from approximately 1 ft to 2 ft. Longitudinal reinforcement ratios ranged from approximately 1.5 to 2.5 percent. The minimum spiral reinforcement ratio was 1.3 percent.

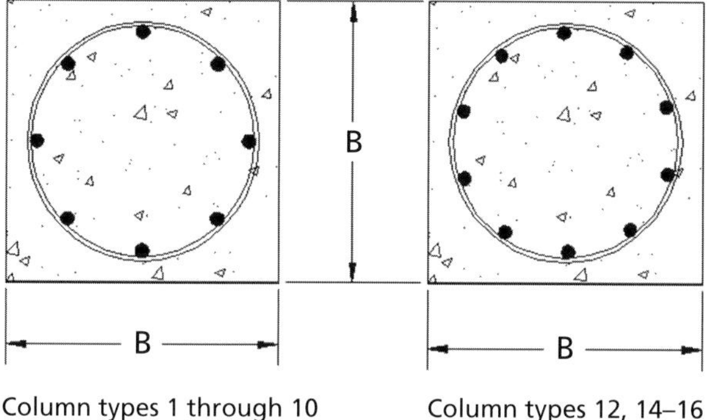

Column types 1 through 10 Column types 12, 14–16

Figure 7.1 Reinforcement arrangement in first-story columns

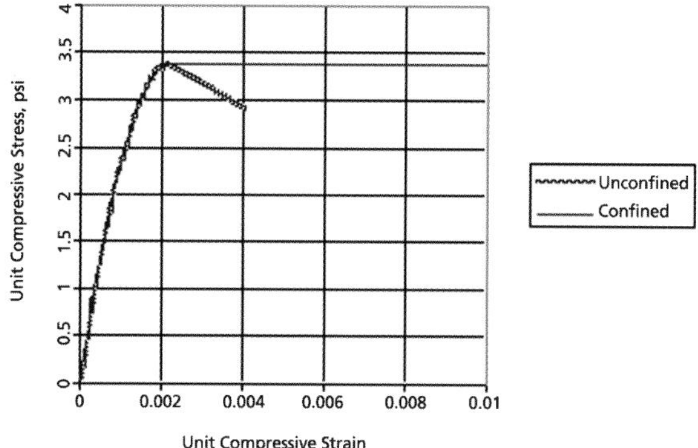

Figure 7.2 Assumed stress-strain curves for confined and unconfined concrete

Moment-curvature relationships for these columns were calculated assuming a mean concrete cylinder strength of 4,000 psi and a yield stress in the longitudinal reinforcement of 45,000 psi. For the concrete stress-strain properties, two different assumptions were made, as shown in figure 7.2. Assumption 1 was used for the gross area of the column treated as a "tied" column and corresponds to unconfined concrete, with the compressive strength of the concrete in the column 85 percent of that in the test cylinder. Assumption 2 was used for the core of the column confined by the spiral reinforcement. For the confined core, the limiting strain was defined to be that corresponding to the fracture of the reinforcement at a unit strain of 0.2. For calculating the relationship between the resisting moment and unit curvature for each type of column, an estimated service load was used reflecting the tributary dead load of the structure. A representative example of the calculated moment-curvature relationships is provided in figure 7.3. The spirally reinforced concrete core had a considerably higher calculated limiting unit curvature capacity than that calculated for the gross section of the column treated as a "tied column." The spiral cores possessed two other important properties not evident in those plots that define only cross-sectional response:

1) The cores enclosed by spiral reinforcement had shear strength higher than the shear corresponding to that associated with the development of the flexural strength of the core under

lateral loading. For the limiting static uniform load corresponding to the critical failure mechanism, the maximum unit shear stress did not exceed three-fourths of the estimated unit shear strength of the core.

2) The longitudinal bars had sufficient anchorage to develop their strengths.

These two properties eliminated the possibility of brittle failure of the cores. Indeed, none of the columns was observed to have failed in shear, and there was no evidence of pull-out of reinforcing bars. The cores and their connections did not unravel under impact. Destroying the column core required tearing it off its supports. The longitudinal reinforcing bars at ends of the severely damaged columns were observed to have fractured after necking, indicating ductile failure.

The plot describing the response of the gross section of the columns (tied columns) refers to a section subjected to flexure with the shell concrete intact and assuming that the shear stresses would not precipitate failure. Had the columns been tied columns—that is, columns without spiral reinforcement confining the core—even the modest unit-curvature limits shown in the figures would not have been attained because shear failure would have preceded development of the yield moments at the critical sections.

The impact effects may be represented as a violent flow through the structure of a "fluid" consisting of aviation fuel and solid fragments. The first-story columns in the path of this rushing fluid mass must have lost their shells immediately on impact. The curves with the higher moment capacities are, in effect, irrelevant for the affected columns. It is very likely that there was never a finite time in which the affected columns responded as tied columns. The column shells must have been scoured off on first contact with the fluid. Bending resistance to the pressure created by the velocity of the fluid must have occurred in the cores only. The limits of the moment-curvature relationships for the column cores shown were based on a nominal fracture strain of 0.2 in the reinforcement in tension. Considering that the axial loads were relatively light in all cases, the curvature limit is more properly based on fracture of the reinforcement. Such a limit, whether it is controlled by limiting strains in the concrete or in the reinforcement, is difficult to determine without directly relevant experimental data because the strain distribution over the column section in the regions of plastic hinging becomes acutely nonlinear at that stage of behavior. The conversion of calculated curvature to rotation is also hampered by the difficulties in defining the deformed geometry in the region of nonlinear response.

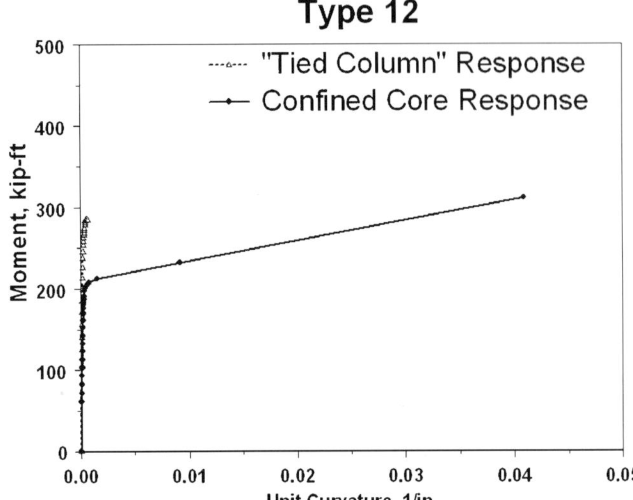

Figure 7.3 Moment-curvature relationship for type 12 column

In keeping with common practice for determining the limiting drift of reinforced-concrete elements, it was assumed that the calculated limiting curvature occurred over a length equal to the core depth. The calculated limiting rotations ranged from 0.2 to 0.5. Accordingly, the spirally reinforced cores of the first-story columns would be expected to tolerate large deflections and still maintain their integrity and absorb energy in bending provided the axial load were transferred to neighboring columns.

Limiting the total energy absorbed by a column to the work done at flexural hinges at the top, bottom, and midheight of the column, the total energy absorbed would be

$$W_{int} = 4M_r \cdot \theta_{limit}$$

where M_r is the resisting moment and θ_{limit} is the limiting concentrated rotation.

Assuming rigid plastic response (recognizing that the response is neither initially rigid nor eventually plastic in the exact sense) and assuming that the impact imparted an initial velocity to the column, the maximum velocity that the column could sustain without disintegration would then be estimated by

$$v_{limit} = (2 \cdot W_{int} / \text{Mass}_{eff})^{1/2}$$

where Mass_{eff} may be taken as one-half the total mass of the column. Evaluation of the above expression for velocity resulted in limiting column initial velocities ranging from approximately 100 to 200 ft/s for the column cores analyzed.

Figure 7.4

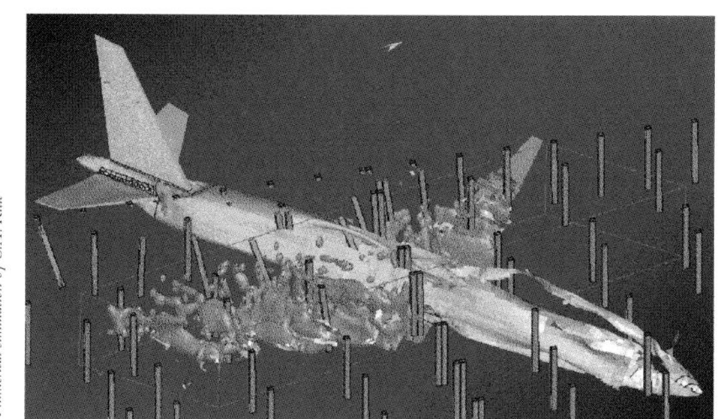

Numerical simulation by S.A. Kilic

Idealized representation of impact on columns

Impacted column

Several numerical simulations of a fluid mass (in this case modeled as aviation fuel) impacting a reinforced-concrete column fixed top and bottom were made by S.A. Kilic in support of the BPS study of the Pentagon. These simulations indicated that the maximum response velocity of the column was comparable to the velocity of the impacting fluid. The conclusion for the facade columns is self-evident. Their maximum response velocities could not have been less than 600 ft/s (vis-à-vis the impact velocity of approximately 780 ft/s). These columns engulfed by the fluid would have been destroyed immediately, however much energy might have been deflected by the facade walls and slabs. The question of interest is whether there was any system to the distribution of the severely damaged columns in the first story.

It is plausible to expect that the energy content of the impacting fluid mass attenuated—as it penetrated the building—as the square of the distance from the point of impact. Recognizing that the debris was not thrown more than a distance of 310 ft and accepting the impact velocity of approximately 780 ft/s, it may be inferred that the velocity of the fluid would have reached a value of approximately 100 ft/s, a velocity that, at a distance approaching 200 ft from the point of impact, most column cores would be expected to resist without disintegration.

There is no question that the progress of the impacting fluid in the structure must have verged on the chaotic. The reasoning in the preceding paragraphs is not presented as a prediction of an orderly process but as a preliminary rationalization of the distribution of severe damage to the spirally reinforced column cores immediately after impact. The important conclusion is that the observed distribution of failed columns does not contradict simple estimates made on the basis of elementary mechanics. There is promise in further analyses of the phenomena observed. The same reasoning would suggest that had the columns in the affected region been tied columns, all would have been destroyed, leading to immediate collapse of a large portion of the building.

A frame from a physics-based simulation of an idealized airplane loaded with fuel impacting a set of spirally reinforced concrete columns (by Hoffmann and Kilic of Purdue University) is shown in figure 7.4. Although completely notional, their analysis senses the deceleration of the airframe as indicated by the buckling of the fuselage. It is also interesting to note that the columns are shown to tear into the airframe but get destroyed by the mass of the fluid in the

Figure 7.5a Column impacted and torn off at supports

Figure 7.5b Ends of fractured reinforcing bars showing necking

wing tanks, events confirmed by the distribution of the debris.

The photograph in figure 7.5a shows a column that had been torn off of its supports. Figure 7.5b is a close-up of the ends of the fractured reinforcing bars. The necking of the reinforcing bars is evidence of the proper performance of the bar anchorages. If energy absorption is a design objective, the evidence suggests that spirally reinforced concrete columns are the right choice.

7.2 LOAD CAPACITY OF FLOOR SYSTEM

The segment of the building that was exposed to heavy impact and fire but survived is bounded by column lines 1 and 11 (figure 7.6a) and AA and O (Ring E exterior wall and AE Drive; figures 2.2, 2.9, and 6.3). In the following text, this segment will be designated as segment P. The small segment to the right of the expansion joint at column line 11 (double columns) includes the segment of the structure that eventually collapsed. In this study, it is taken to be bounded by column lines 11, 17, AA, and D. This segment will be designated segment Q (figure 7.6b).

7.2.1 FLEXURAL STRENGTH OF FLOOR SYSTEM WITH MISSING COLUMNS

The limiting flexural strength of the floor system was determined to establish a measure of the strength of the system. The investigation was made not to find out how much load the system would carry but to determine the quality of the construction by using the limiting flexural capacity as an index value. Flexural capacities at critical sections were determined, ignoring the effect of compression reinforcement. It was assumed at the support sections that the tensile reinforcement in the slab acting as a flange was effective in resisting flexure. The width of the flange was defined to be equal to the clear depth of the beam below the slab. Considering the moment gradient along the span at sections where flexural yield was expected, the effective stress in the tensile reinforcement was assumed to be $^5/_4$ times the yield stress at room temperature.

Unit resisting loads corresponding to the development of flexural failure mechanisms with yield lines paralleling the column centerlines (that is, yielding the girders with all original columns in place) were determined for segments P and Q.

The minimum unit yield load calculated was more than 1,300 psf. The calculations were repeated, assuming that entire rows of columns were missing, resulting in a span length of 40 ft center to center of the columns. For those conditions, the unit load calculated was not less than 300 psf. Similar results were obtained assuming a row of interior columns along lettered column lines missing, which results in a longer span for the beams. The calculated probable value of 1,300 psf does not refer to the actual capacity of the structure, as other modes of failure might have governed before this load could be achieved. But it does attest to the impressive intrinsic strength of the floor system and explains, along with the observation that the bottom bars were lapped at column lines, why the structure could tolerate losses of columns. Simplified yield line analyses were also performed for areas within segment P that were missing several columns. Figure 7.6a shows two such areas, P1 and P2. The two areas are shown divided because the light well wall above (see figure 2.10, section 2) provides a stiff and strong support, and the second-story columns were able to act as hangers because the column vertical reinforcement was sufficiently well developed (having the lap enclosed within a spiral may have been a factor). The calculated capacities of the floor systems within P1 and P2 were both over 350 psf, more than twice the dead load. The light well wall was also analyzed for the loads that the area P1 and P2 mechanisms would deliver to it, and this analysis showed that some support from the two significantly impaired columns at 9 and 11 F would have been required to prevent a failure. The spiral reinforcement in those two columns must have been a key factor in preventing a widespread collapse.

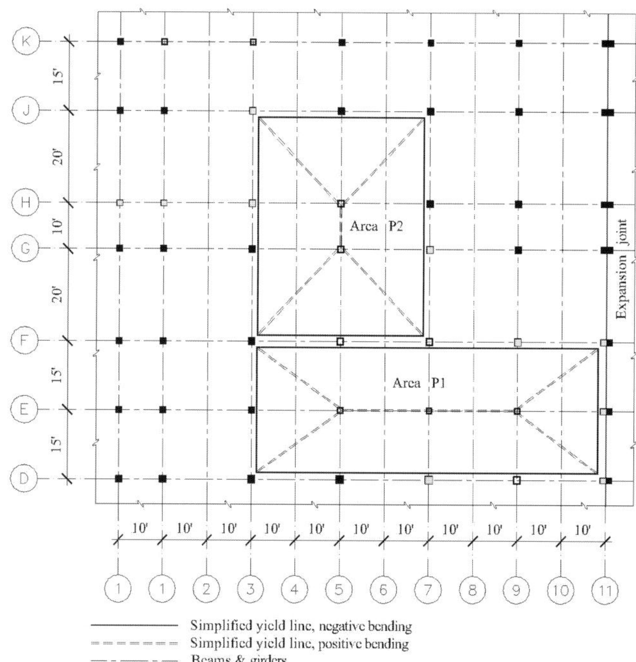

Figure 7.6a Yield line analysis of area with missing columns

7.2.2 STRENGTH OF FLOOR SYSTEM IN COLLAPSE AREA

Figure 7.6b depicts the segment of the building between column lines 11 and 17 and AA and D. To obtain another perspective of the intrinsic strength of the floor system, its capacity to resist load at room temperature was estimated by assuming that all columns at the intersections of column lines 12 through 16 and AA, A, B, and C were lost. The assumed locations of the negative- and positive-moment yield lines are shown in the figure. The floor-system edge along line 11—the location of the expansion joint—was assumed to be unsupported. The support along line AA was considered to be provided by the facade wall that had not been destroyed by the impact. (An example of a similar phenomenon at a different building is shown in figure 7.7.)

The calculated capacity for the failure condition shown ideally in figure 7.6b was approximately 160 psf (with the dimension x set at 35 ft, corresponding to the minimum calculated yield load). For the assumed boundary, material, and support conditions, the floor system at level two would have been able to support itself over the assumed unsupported area. It is plausible to assume that the columns at the intersections of line AA with lines 11 through 16 were not functional at the time the photograph that constitutes figure 3.8 (section 3) was taken. Photographs taken immediately after the impact (figures 3.8 and 3.9, section 3) indicate that segment Q might have derived some support from segment P before collapsing. The partial support might have provided a collateral

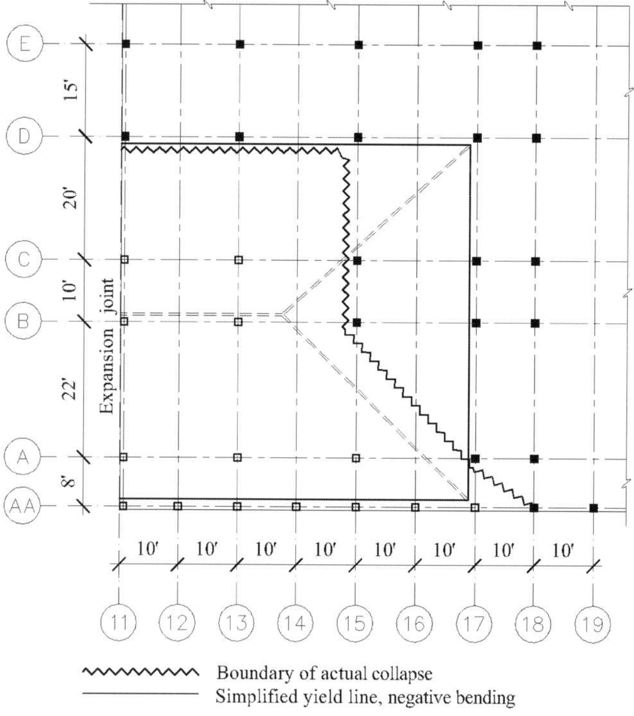

Figure 7.6b Yield line analysis of collapse area, segment Q

mechanism for resisting the overall gravity loads on segment Q considered as a building block, but the floor system would still have had to be able to carry itself. The eventual collapse of this section is attributed to the effects of heat from the fire or high-strain creep, possibly exacerbated by the water pumped into the structure to quell the fire.

7.3 THERMAL RESPONSE OF COLUMNS AND GIRDERS

Prior to the collapse of portions of the structural system in Wedge 1 of Ring E, which occurred approximately 20 minutes after the impact of the aircraft, the fire that was first ignited by the ejected jet A fuel had transitioned from the growth stage and become a ventilation-controlled "fully developed" or "postflashover" fire. This is evidenced in figure 3.8 (see section 3), which was taken prior to the collapse (within the first half-hour following the aircraft impact) and shows the flames projecting from the windows. In a ventilation-controlled postflashover fire, the flames typically project from windows and openings because there is insufficient air in the burning rooms to allow all the combustible gases to burn within the rooms.

Estimation of the fire intensity—that is, maximum temperatures and time-temperature characteristics—of postflashover fires is important in understanding the effect of fire on exposed structural elements. However, the accuracy of such estimation depends on a correct estimation of the fire fuel load (hydrocarbon-based building and aircraft contents and jet A fuel) and the ventilation factor. This cannot be done with a high degree of exactness even in a typical building fire. In the case of the Pentagon attack, it is further complicated by the lack of complete knowledge of the available fuel load (besides the ejected jet A fuel) and by the unconventional ventilation factor.

In the sections that follow, the lower bound of the maximum temperature and the time-temperature curve of the Pentagon fire will be estimated using widely cited published data as well as damage observations made by the BPS team. The assumptions used to establish the lower bound of the real fire are also outlined.

7.3.1. LOADING

The fire intensity can be estimated if the fire fuel load, e_f (shown in MJ/m^2 in figure 7.8), and the ventilation factor, F_v (shown in $m^{1/2}$ in figure 7.8), are known. The maximum fuel capacity listed for the Boeing 757-200 is 11,275 gal (42,680 L) *(www.boeing.com)*. According to information provided by the National Transportation Safety Board, the aircraft had on board about 5,300 gal (20,200 L) of jet A fuel, or approximately 36,200 lb (16,000 kg) of fuel based on the density of 6.8 lb/gal (0.79 g/cm^3), at the time of impact. Based on images captured by the Pentagon security camera, which showed the aircraft approaching and the subsequent explosion and fireball, it is estimated that about 4,900 lb (2,200 kg) of jet fuel was involved in the prompt fire and was consumed at the time of impact outside the building. This leaves about 30,400 lb (13,800 kg) as the estimated mass, M, of the jet A fuel that entered the building and contributed to the fire fuel load within the building.

The net calorific value or heat of combustion—that is, the amount of heat released during complete combustion of a unit mass of fuel, ΔH_c—measured for jet A fuel is 18,916.6 Btu/lb (44 MJ/kg). Thus, the maximum possible energy, E, that could have been released inside the building by the complete burning of 30,400 lb (13,800 kg) of jet A fuel is

Mike Johns

Figure 7.7 Reinforced-concrete frame with masonry filler walls acting as a vertical diaphragm to compensate for column loss

$$E = M \cdot \Delta H_c = 30{,}400 \times 18{,}916.6 = 575{,}064{,}488 \text{ Btu } (607{,}200 \text{ MJ})$$

It is assumed that the fuel was initially contained within the first floor, in a "room" bounded by the path of damage caused by the impact of the airplane (shaded gray in figure 7.9). The estimated total surface area, A_t (floor, ceiling, and bounding walls including windows and openings), of the room is about 36,597 sq ft (3,400 m^2). The fire fuel load, $e_{f,a}$, contributed by the available jet A fuel alone can be computed as

$$e_{f,a} = E/A_t = 575{,}064{,}488/36{,}597$$
$$= 15{,}713 \text{ Btu/sq ft } (178 \text{ MJ/m}^2)$$

As indicated, within the first half an hour of the aircraft impact, the fire had become fully developed within some compartments of the Pentagon. This means combustible building and aircraft contents had begun to burn and therefore contributed to the fire fuel load. The exact fire fuel load contributed by the building and aircraft contents, $e_{f,b}$, is not known because of insufficient information on the type of occupancy in this particular section of the Pentagon. However, a lower-bound estimate can be made using data recommended by the International Council for Research and Innovation in Building and Construction, or CIB, which lists average fuel loads for different types of building occupancy (International Council 1986). Occupancy types that might be similar to those of the Pentagon are given in table 7.1a.

It is assumed that the type of occupancy of the Pentagon is such that the fire fuel load of its building and aircraft contents is equivalent to the lowest value of the four CIB office types of occupancy given in table 7.1a. Also, since the CIB-recommended fuel loads are for design purposes, it is believed that they include the safety factor, the magnitude of which is unfortunately not known. Thus, a conservative safety factor of 2 can be assumed in the CIB recommendation. The lower bound of the fire fuel load contributed by the building and aircraft contents, $e_{f,b}$, can then be estimated to be about 17,611 Btu/sq ft (200 MJ/m^2). The combined total fire fuel load, e_f, can then be estimated to be about 33,325 Btu/sq ft (378 MJ/m^2).

The ventilation factor, F_v, can be computed as

$$F_v = A_v (h^{1/2})/A_t$$

where A_v and h are respectively the area and height of the "room" opening. The room opening in this case is estimated to be about 75 percent of the total area of the building elevation along column line AA that is limited to the first story and bounded between column lines 8 and 19. The 75 percent area accounted for the existing windows and the opening created by the impact of the airplane. The total surface area in the first story between column lines 8 and 19 is about 1,098 sq ft (102 m^2) based on a height, h, of 10 ft (3.05 m). Thus, A_v can be estimated as

$$A_v = 0.75 \times 1{,}098 = 824 \text{ sq ft } (77 \text{ m}^2)$$

and

$$F_v = 824(10^{1/2})/36{,}597 = 0.071 \text{ ft}^{1/2} \ (0.04 \text{ m}^{1/2})$$

The time-temperature curves for different fuel loads, e_f, and ventilation factors, F_v, produced by Magnusson and Thelandersson (1970) are widely used for estimating real fire exposure. The curves for a ventilation factor, F_v, of 0.04 m$^{1/2}$, and fuel loads ranging from 2,200 to 44,027 Btu/sq ft (25 to 500 MJ/m^2) are shown

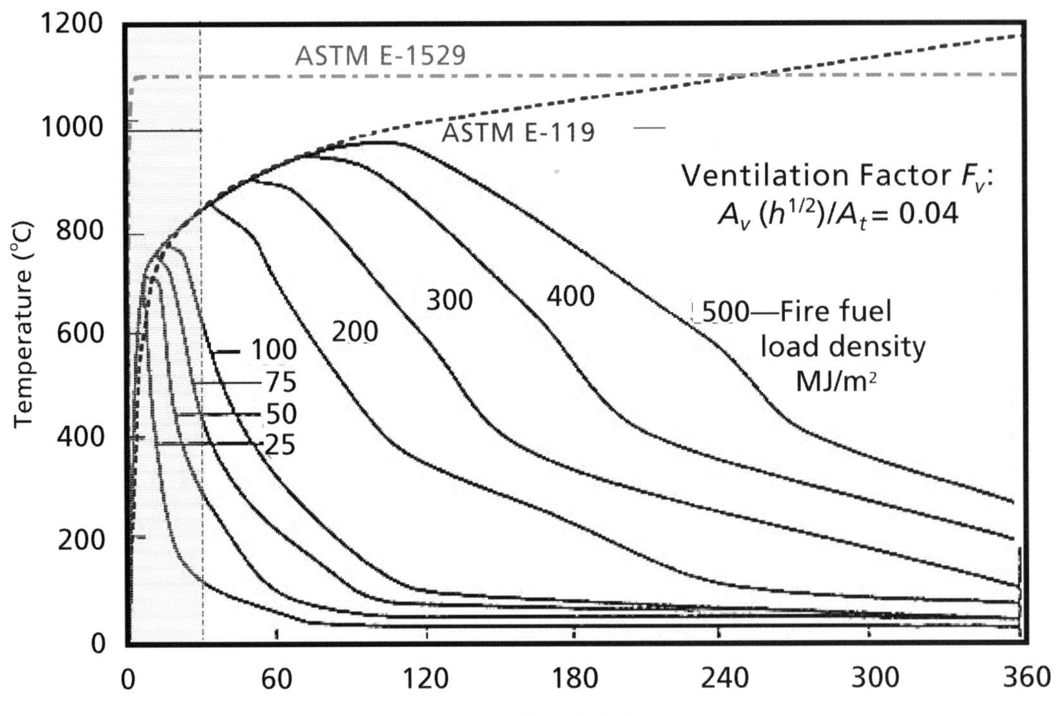

Figure 7.8 Time-temperature curves for different fuel loads (Magnusson and Thelandersson 1970)

in figure 7.8. Since the estimated lower bound of the fire fuel load is 33,325 Btu/sq ft (378 MJ/m^2), the time-temperature curve for the fire can be estimated to be between the curves for 26,416 Btu/sq ft (300 MJ/m^2) and 35,222 Btu/sq ft (400 MJ/m^2) in this figure, which means an estimated maximum temperature of close to 1,560°F (850°C) in a little more than 30 minutes.

It should be noted that the estimated time-temperature curves for all fire fuel loads in this figure have the same initial rate of temperature rise (the first 10 minutes of the fire), and this initial rate of temperature rise is higher than that prescribed for standard fire ASTM E-119 but lower than that of standard hydrocarbon pool fire ASTM E-1529. Similarly, within the first half-hour of the fire (prior to collapse) the temperature of the estimated fire was slightly higher than the ASTM E-119 temperature but lower than the temperature prescribed by ASTM E-1529. The shaded portion of figure 7.8 highlights the temperature profile of the first half-hour of the estimated fire.

Table 7.1a

Type of Occupancy	Fuel Load Btu/sq ft (MJ/m^2)
Administration	70,444 (800)
Data Processing	35,222 (400)
Institution Building	44,027 (500)
Engineering Office	52,833 (600)

7.3.2 RESPONSE

The Pentagon building comprises several reinforced-concrete structural frame systems, which are frame constructions consisting of reinforced-concrete beams, columns, and slabs. The impacted section of the Pentagon is located in Wedge 1 and is composed of two reinforced-concrete structural systems separated by an expansion joint at column line 11. According to eyewitness accounts of the Pentagon attack, the structure survived the initial aircraft impact—that is, it did not collapse. However, portions of the structure system in Wedge 1/Ring E, south of the expansion joint at column line 11 and directly above where the aircraft impacted, collapsed approximately 20 minutes after the impact and exposure to the ensuing fire. Figure 7.9 shows the plan view of the affected area of the Pentagon and the collapsed section. As shown in this figure, collapse was limited to Ring E, adjacent to the expansion joint at column line 11, and at a section that had the largest unsupported floor area because of loss of first-story columns. Other damaged sections in rings D and E (to the north of the collapsed section and

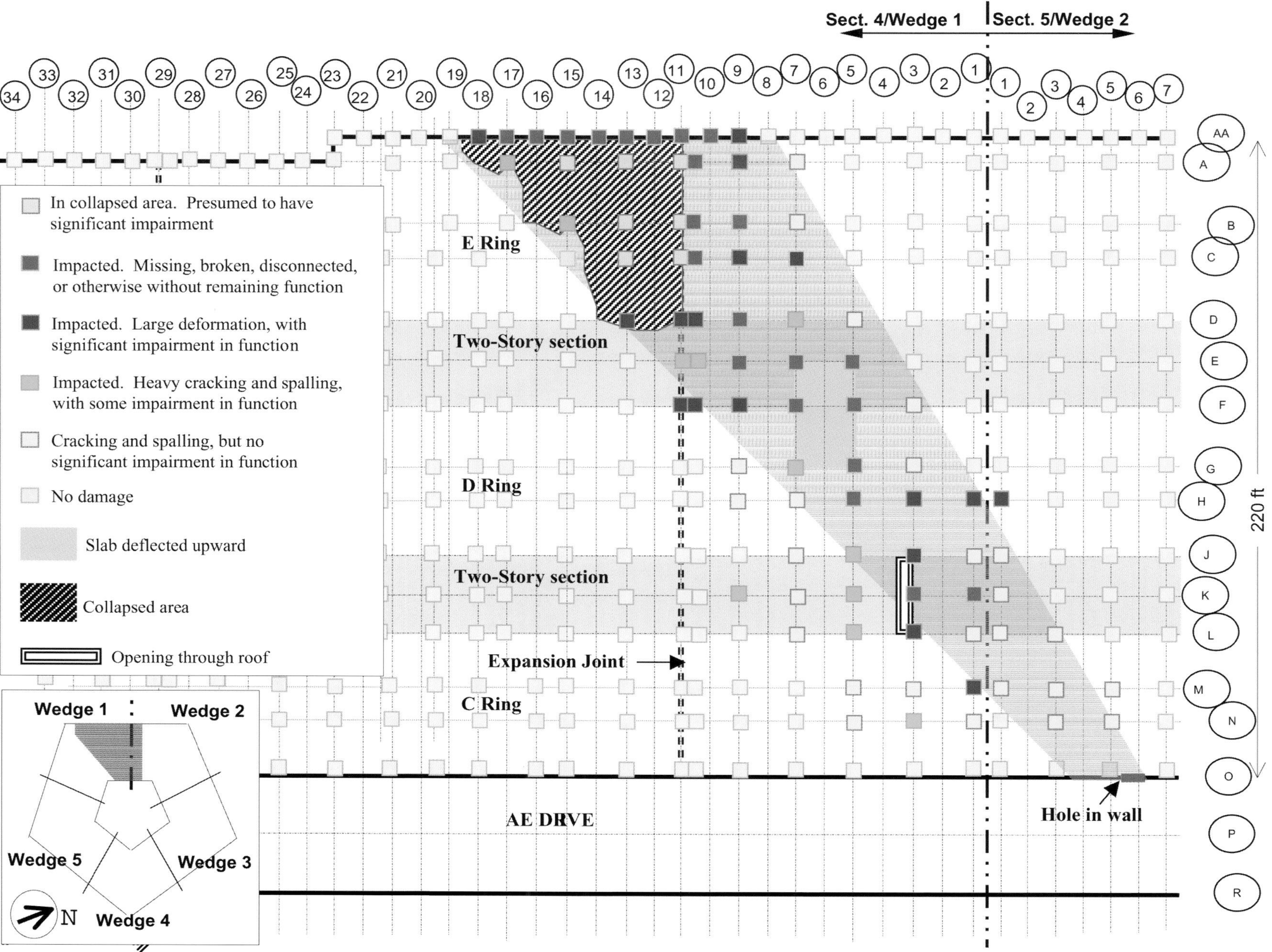

Figure 7.9 Path of ejected fuel in the first story

14 in.
Undamaged Column
Damaged Column
14 in.
Undamaged Beam
Damaged Beam

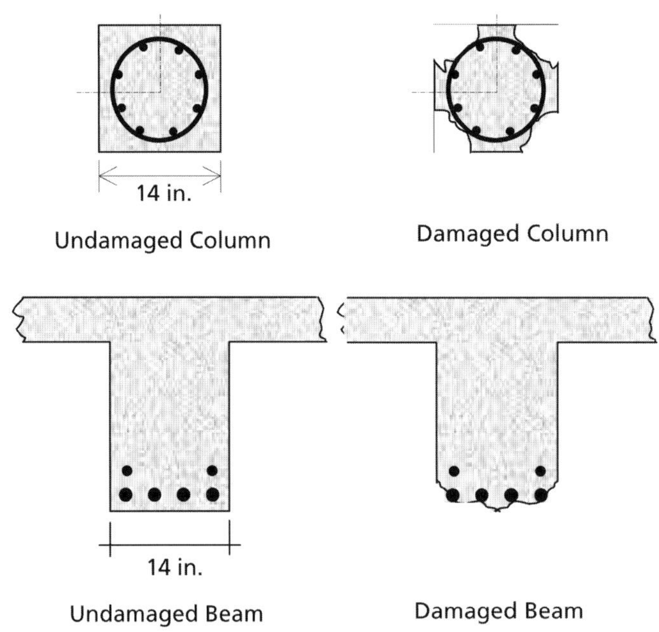

Figure 7.10 Cross-sectional properties of beams and columns analyzed

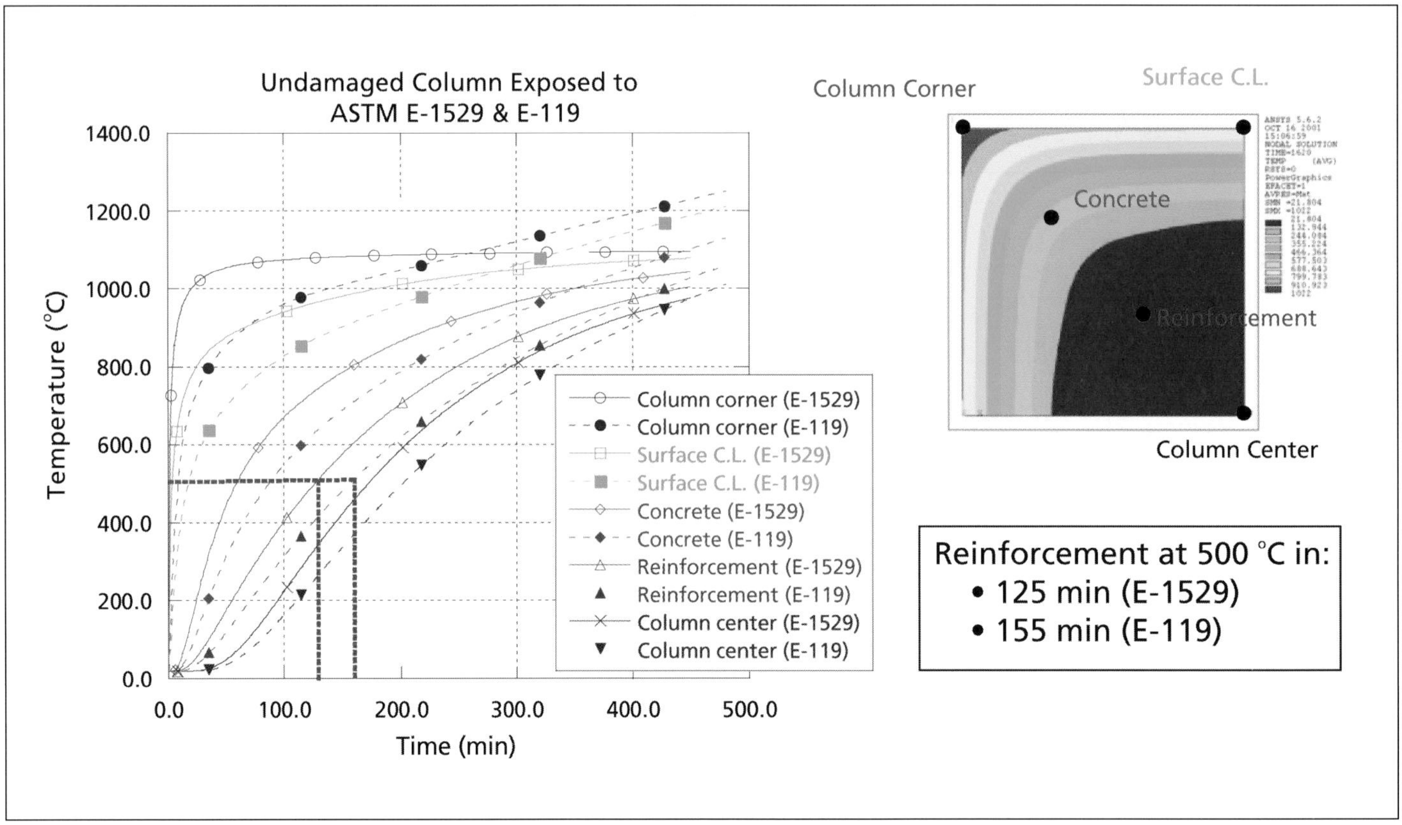

Undamaged Column Exposed to ASTM E-1529 & E-119
Temperature (°C)
Time (min)
0.0
200.0
400.0
600.0
800.0
1000.0
1200.0
1400.0
0.0
100.0
200.0
300.0
400.0
500.0
Column corner (E-1529)
Column corner (E-119)
Surface C.L. (E-1529)
Surface C.L. (E-119)
Concrete (E-1529)
Concrete (E-119)
Reinforcement (E-1529)
Reinforcement (E-119)
Column center (E-1529)
Column center (E-119)
Column Corner
Surface C.L.
Concrete
Reinforcement
Column Center
Reinforcement at 500 °C in:
• 125 min (E-1529)
• 155 min (E-119)

Figure 7.11 Temperature development in undamaged column

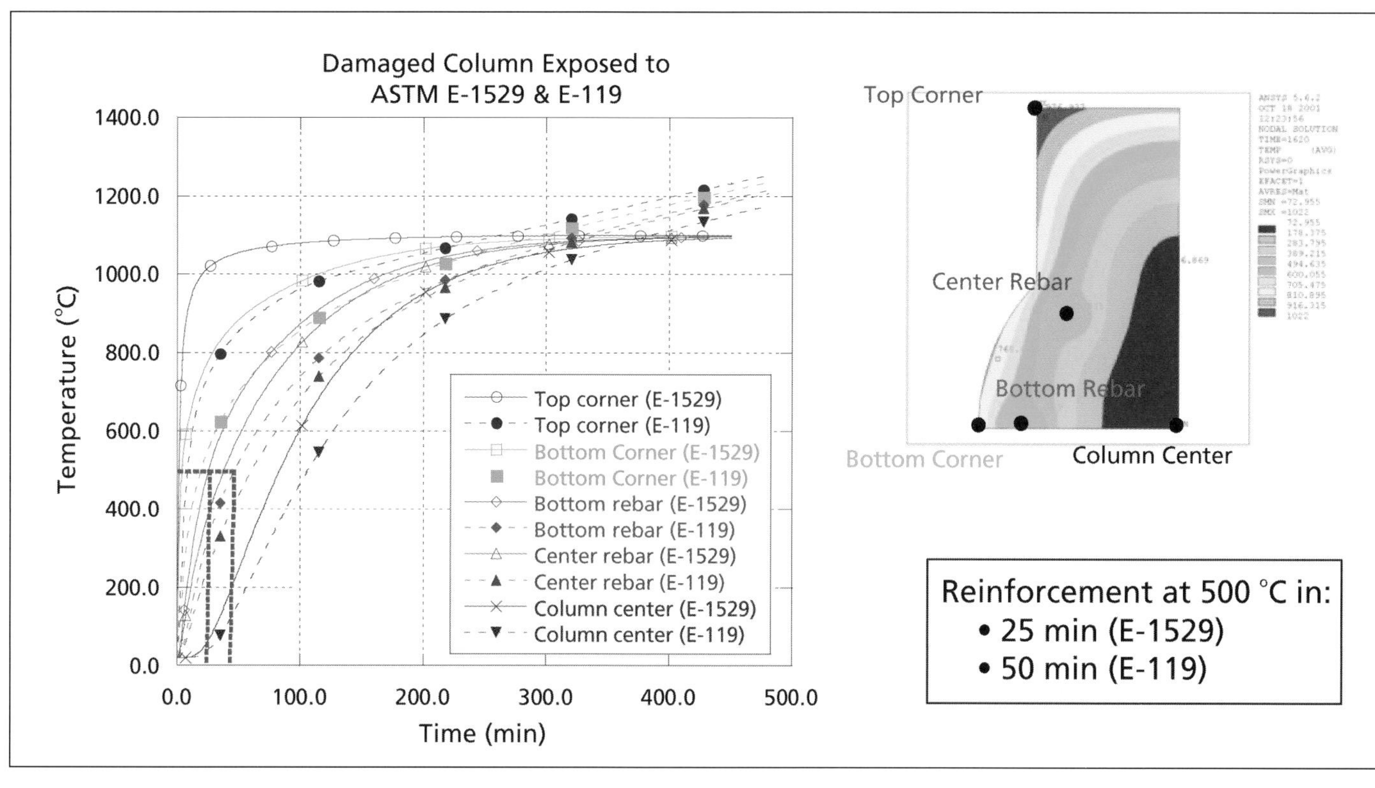

Figure 7.12 Temperature development in damaged column

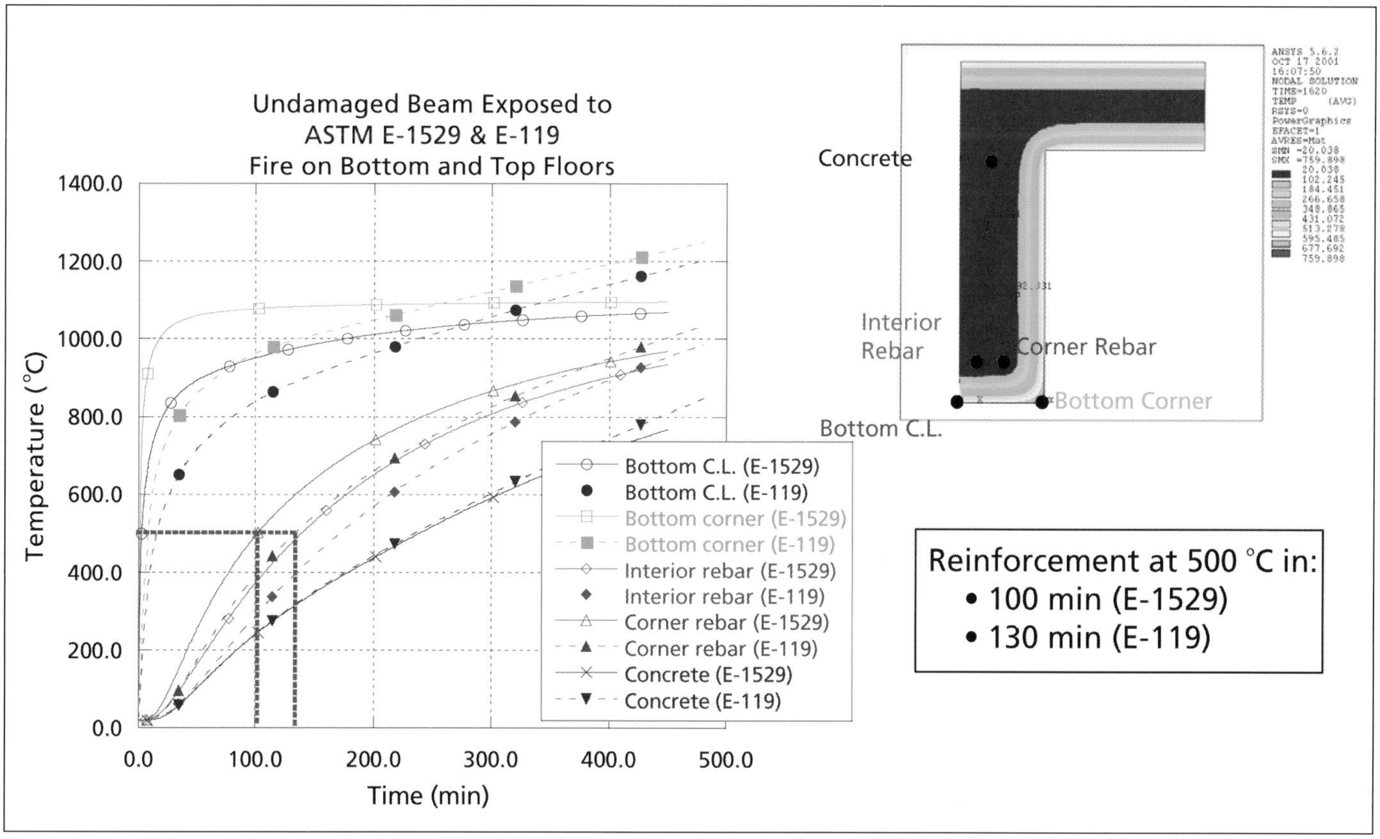

Figure 7.13 Temperature development in undamaged beam

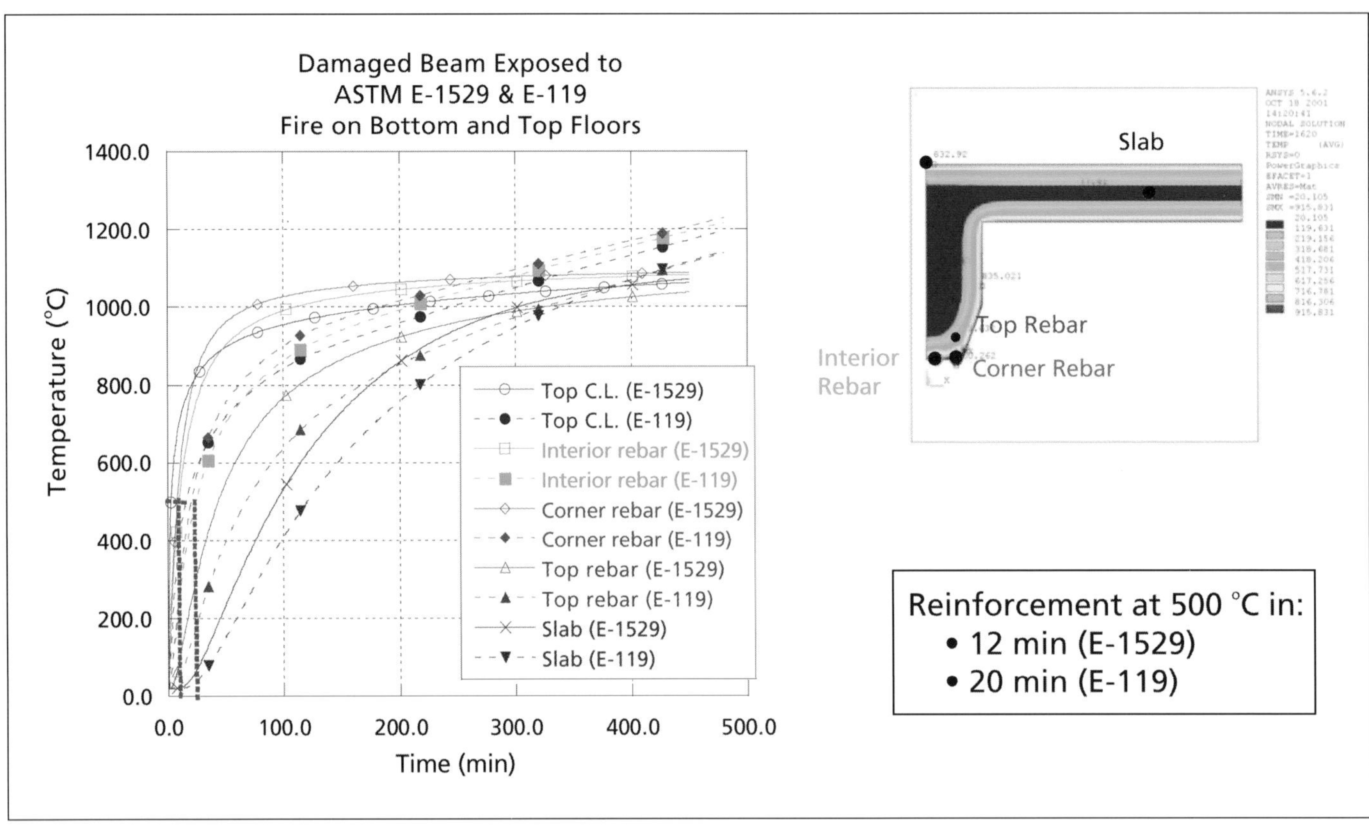

Figure 7.14 Temperature development in damaged beam

the expansion joint) apparently had smaller unsupported floor areas and survived both the impact and the fire.

While many structural members (columns and main girders) in the impact zone were destroyed by the aircraft impact (broken, disconnected, or with large deformation [damage classifications 4 (column 9B) and 5 (columns 9A and 9AA) described in section 6; see figures 6.2–6.6, section 6], the degree of damage to other members in and adjacent to the impact zone varied, and in some instances was limited to loss of concrete cover only (example column 15B; see figure 5.18, section 5). Under normal ambient temperature, the loss of concrete cover in itself does not significantly affect the structural capacity of the structural system. However, when the reinforcement of stripped members is exposed directly to fire, as it was in this case, the load-carrying capacity of individual structural members, and therefore of the entire system, can quickly deteriorate because of high temperature exposure. This loss in capacity could lead to premature collapse of the entire structural system, particularly in one with such severe mechanical damage.

As established in the above section, the time-temperature curve for the estimated real fire exposure for some compartments in the Pentagon is between those prescribed for the standard building fire (ASTM E-119 and ISO 834) and the standard hydrocarbon pool fire (ASTM E-1529), in terms of both maximum temperature and rate of temperature rise in the first hour of the fire. Thus, the two standard fires, ASTM E-119 and ASTM E-1529, can be considered the lower and upper bounds of the actual fire condition in the first hour and will be used as the fire conditions for purposes of thermal analysis. The time-temperature relationships for the standard fire exposures are shown in figure 7.8.

A typical column (type 5, 14 by 14 in.) and main girder (14 by 20 in.) were subjected to thermal analysis. The column and girder were analyzed in two conditions: undamaged and damaged. Damaged columns and girders were members that had lost a portion of the concrete cover for the main reinforcement as a consequence of the impact but were not broken. This damaged condition was simulated by removing the concrete finite elements that modeled the reinforcement cover at one corner of the column and girder, thereby exposing the reinforcing spiral and the longitudinal rebar directly to heat. Figure 7.10 shows the dimensions and reinforcement details of the column and girder analyzed.

The transfer of heat from the standard fire exposures to the column and girder was by way of forced convection with an assumed heat transfer coefficient of 567.8 Btu/(h·ft^2·°F) (100 W/m^2·K). The concrete and reinforcement were discretized into two-dimensional, four-noded solid thermal elements with one degree of freedom (temperature) at each nodal point.

The thermal material properties set forth in table 7.1b were used.

In the analysis, all interior finishes that provided added fire protection to the columns and girder were assumed to have been stripped off by the impact. Figures 7.11 through 7.14 summarize the results of this analysis.

The procedure given in Eurocode 2—part 1-2 ("Structural Fire Design")—was also used to estimate the fire endurance of an undamaged column and girder. As shown in table 7.2, the fire endurances of the undamaged siliceous aggregate concrete column and girder are estimated to be greater than respectively 90 and 120 minutes under the ISO 834 fire exposure. This represents the lower bound in fire endurance of the undamaged column and girder. However, once portions of the concrete cover for the reinforcement have been stripped off, the structural capacities of the damaged columns and girders can quickly be compromised, as reflected in the shortened fire endurance times obtained from the thermal analysis shown in figures 7.11 through 7.14 and listed in table 7.2. The fire endurance times for the column and beam are determined as the times for the main reinforcing steel to reach critical temperature (932°F [500°C], impending yielding temperature of steel reinforcement as defined in the Eurocode). For the column, the fire endurance time when damaged ranges from 25 minutes (based on the upper-bound fire ASTM E-1529) to 50 minutes (based on the lower-bound fire ASTM E-119). For the girder, the fire endurance time when damaged ranges from 12 minutes (ASTM E-1529) to 20 minutes (ASTM E-119). These endurance times compare well with the observed time to collapse after the initiation of the fire.

Table 7.1b

Property	Concrete	Steel
Density	150 lb/cu ft (2,400 kg/m^3)	493 lb/cu ft (7,900 kg/m^3)
Thermal conductivity	908.6 Btu·in./(s·ft^2·°F) (1.75 W/m·K)	26,948 Btu·in./(s·ft^2·°F) (51.9 W/m·K)
Specific heat	4,186,800 Btu/(lb·°F) (1,000 J/kg·K)	2,034,785 Btu/(lb·°F) (486 J/kg·K)
Thermal diffusivity	7.85×10^{-6} ft^2/s (7.29×10^{-7} m^2/s)	1.45×10^{-4} ft^2/s (1.35×10^{-5} m^2/s)

Table 7.2

Member	Condition	Fire Exposure	Fire Endurance Time Based on Eurocode (minutes)	Time to Critical Temperature Based on Thermal Analysis (minutes)
Column	Undamaged	E-119	≥90	155
Column	Undamaged	E-1529	N/A	125
Column	Stripped to spiral	E-119	N/A	50
Column	Stripped to spiral	E-1529	N/A	25
Girder	Undamaged	E-119	≥120	130
Girder	Undamaged	E-1529	N/A	100
Girder	Stripped to rebar	E-119	N/A	20
Girder	Stripped to rebar	E-1529	N/A	12

8. FINDINGS

Through observations at the crash site and approximate analyses, the team determined that the direct impact of the aircraft destroyed the load capacity of about 30 first-floor columns and significantly impaired that of about 20 others along a diagonal path that extended along a swath that was approximately 75 ft wide by 230 ft long through the first floor. This impact may also have destroyed the load capacity of about six second-floor columns adjacent to the exterior wall. While the impact scoured the cover of around 30 other columns, their spiral reinforcement conspicuously preserved some of their load capacity. The impact further destroyed the load capacity of the second-floor system adjoining the exterior wall.

The subsequent fire fed by the aircraft fuel, the aircraft contents, and the building contents caused damage throughout a very large area of the first story, a significant area of the second, a small part of the third, and only in the stairwells above. This fire caused serious spalling of the reinforced-concrete frame only in a few, small, isolated areas on the first and second stories. Subsequent petrographic examination showed more widespread heat damage to the concrete.

Despite the extensive column damage on the first floor, the collapse of the floors above was extremely limited. Frame and yield-line analyses attribute this life-saving response to the following factors:

- Redundant and alternative load paths of the beam and girder framing system;
- Short spans between columns;
- Substantial continuity of beam and girder bottom reinforcement through the supports;
- Design for 150 psf warehouse live load in excess of service load;
- Significant residual load capacity of damaged spirally reinforced columns;
- Ability of the exterior walls to act as transfer girders.

An area covering approximately 50 by 60 ft of the upper floors above the point of impact did collapse approximately 20 minutes after the impact. Thermal analyses indicate that the deleterious effect of the fire on the structural frame, together with impact damage that removed protective materials and compromised strength initially, was the likely cause of the limited collapse in this region.

9. RECOMMENDATIONS

9.1 DESIGN AND CONSTRUCTION

The Pentagon's structural performance during and immediately following the September 11 crash has validated measures to reduce collapse from severely abnormal loads. These include the following features in the structural system:

- Continuity, as in the extension of bottom beam reinforcement through the girders and bottom girder reinforcement through the columns;
- Redundancy, as in the two-way beam and girder system;
- Energy-absorbing capacity, as in the spirally reinforced columns;
- Reserve strength, as provided by the original design for live load in excess of service.

These practices are examples of details that should be considered in the design and construction of structures required to resist progressive collapse.

9.2 RESEARCH AND DEVELOPMENT

The Pentagon crash supports the need for research and development in progressive collapse and extreme lateral column response. The following topics are of particular interest:

Consolidation of information on prevention of progressive collapse: Much has been written on means to prevent progressive collapse, but little detailed guidance has been incorporated into the building codes in general use in the United States. There should be a focused effort to accumulate research and practical experience in the area of structural robustness so that an authoritative guide can be prepared that will be useful to the design community.

Influence of extreme column deformations on load-carrying capacity: The columns in the Pentagon deformed laterally to several times their diameter. In this highly plastic, postfailure state they continued to function as structural elements. Research should be performed to determine the load-carrying capacity of columns and other structural elements once they have been deformed beyond their maximum load-carrying state and are in the range of declining strength.

Influence of extreme column deformations on loads within a statically indeterminate structure: Once the columns in the Pentagon deformed laterally beyond a certain amount, most certainly they began to pull down on the structure above, acting as a catenary. Under this circumstance, the columns placed additional demand on the adjacent structure, at least for the brief time that it experienced the lateral load that caused the horizontal displacement. Research should be conducted to understand the implications of this short-term load on the survivability of structures.

Energy-absorbing capacity of reinforced-concrete elements: The columns of the Pentagon absorbed energy as they deformed at a very high strain rate. Research should be conducted to understand the energy-absorbing capacity of concrete elements when they are subjected to impact and impulse loads that result in large deformation.

Ability of a structure to withstand extreme impact: The data collected and such other data as may be available or can be generated on this subject should be studied to extract information useful to engineers charged with designing buildings so as to reduce risks caused by extreme impacts that can induce extensive damage.

REFERENCES

Arlington County, Virginia. 2002. *After-action report on the response to the September 11 terrorist attack on the Pentagon.*

Department of Transportation. Federal Aviation Administration. September 1987. *Summary report—Full-scale transport controlled impact demonstration project* (DOT/FAA/CT-87/10). FAA Technical Center, Atlantic City International Airport, New Jersey.

Hool, George A., and Harry E. Pulver. 1937. *Reinforced concrete construction. Vol. 1 of Metal principles.* 4th ed. New York: McGraw-Hill Book Company.

International Council for Research and Innovation in Building and Construction, CIB-W14. 1986. Design guide—Structural fire safety. *Fire Safety Journal* 10 (2): 75–138.

Magnusson, S.E., and Thelandersson, S. 1970. Temperature-time curves of complete process of fire development: Theoretical study of wood fuel fires in enclosed spaces. *Acta Polytechnica Scandinavica* (Civil Engineering and Building Construction Series 65).

Phan, L.T., N.J. Carino, D. Duthinh, and E. Garboczi, eds. 1997. *International workshop on fire performance of high-strength concrete* (Special Publication SP 919). Gaithersburg, Maryland: National Institute of Standards and Technology.

Shirley, G., R.G. Burg, and A.E. Fiorato. 1987. Chicago Committee on High-Rise Buildings. *Fire endurance of high-strength concrete slabs.* Report no. 11.

APPENDIX A

ACKNOWLEDGMENTS

The contributions made by the following persons to the building performance study of the Pentagon are gratefully acknowledged.

Report Review Task Committee
Technical Activities Division
Structural Engineering Institute

R. Shankar Nair, Ph.D., P.E.
Senior Vice President
Teng Associates, Inc.
205 North Michigan Avenue, Suite 3600
Chicago, IL 60601

William McGuire, P.E.
Professor of Civil Engineering, Emeritus
School of Civil Engineering
Hollister Hall
Cornell University
Ithaca, NY 14853

Lawrence Griffis, P.E.
President, Structures Division
Walter P. Moore & Associates, Inc.
1221 South Mopac Expressway, Suite 355
Austin, TX 78746

Robert J. McNamara, P.E.
President
McNamara/Salvia, Inc.
160 Federal Street, 16th Floor
Boston, MA 02110

Theodore V. Galambos, Ph.D., P.E.
Emeritus Professor of Structural Engineering
University of Minnesota
Department of Civil Engineering
500 Pillsbury Drive, S.E.
Minneapolis, MN 55455

David Peraza, P.E.
Vice President
LZA Technology
641 Avenue of the Americas
New York, NY 10011

Delbert Boring, P.E.
Senior Director, Construction
American Iron and Steel Institute
1101 17th Street N.W., Suite 1300
Washington, DC 20036

John Hanson, Ph.D., P.E.
Affiliated Consultant (Formerly President)
Wiss, Janney, Elstner Associates, Inc.
330 Pfingsten Road
Northbrook, IL 60062

Reidar Bjorhovde, Ph.D., P.E
President
The Bjorhovde Group
5880 East Territory Avenue
Tucson, AZ 85750

Robert Ratay, Ph.D., P.E.
Consulting Engineer
198 Rockwood Road
Manhasset, NY 11030

Boeing
Rodney Dreisbach
David Grubb
Jack F. McGuire

Simpson Gumpertz & Heger, Inc.
Raffi K. Batalian, Graphics support
Gene G. LeBlanc, Lead illustrator
Kimberly J. Vierstra, Support illustrator
Matthew C. Wagner, Engineering

U.S. Army Engineer Research and Development Center
Richard C. Dove, Ph.D., P.E., Research Engineer
Robert L. Hall, Ph.D., Division Chief
Thomas R. Slawson, Ph.D., P.E., Research Engineer

J.R. Harris & Company
Donald Carroll, P.E., Engineering analysis
William Edmands, Figure preparation

HSMM Architects Engineers Planners
Mike Biscotte, P.E., Structural Engineer

WCMH
Shannon Harris, Executive Producer, video support

USAR Engineers
Dean Tills, Photographs and observations

Purdue University
Information Technology
C. Hoffmann, Ph.D., Professor of Computer Science
S.A. Kilic, Ph.D., Visiting Scholar in Civil Engineering

National Institute of Standards and Technology
Monica A. Starnes, Co-op Student (MIT Doctoral Candidate)

National Transportation Safety Board
James Ritter

Construction Technology Laboratory
W. Gene Corley, Ph.D., P.E.

National Science Foundation
Engineering Directorate
Support for the Simulation Project

Structural Engineering Institute
American Society of Civil Engineers
James A. Rossberg, P.E., Director
Mary Ellen Saville, Administrator

Publications Division
American Society of Civil Engineers
Anne Elizabeth Powell, Editor in Chief
Jan Hilton, Art Director
Chris Ralston, Technical Editor

APPENDIX B

COLUMN LABEL	DESCRIPTION OF DAMAGE	PHOTO (IF AVAILABLE)
9AA	• Bowed • Stripped to spiral reinforcement • 5 to 6 inches out of plumb • Shored	
10AA	• Missing • Remaining upper section connected to girder is stripped to spiral reinforcement • Shored	
11AA	• Missing • Remnants of 8 #8 deformed bars visible at ceiling • Shored	
12AA, 13AA,14AA,15AA, 16AA, 17AA	• Missing	

COLUMN LABEL	DESCRIPTION OF DAMAGE	PHOTO (IF AVAILABLE)
18AA	• Bowed • Concrete completely missing at top • Spalling present from midheight to top • Shored	
11A	• Missing • Shored	
13A,15A	• Missing	
17A	• Stripped to spiral reinforcement at top of column • Shored	

COLUMN LABEL	DESCRIPTION OF DAMAGE	PHOTO (IF AVAILABLE)
7B	• Split by vertical crack that looks typically like thermal crack • Shored	
9B	• Bowed • Stripped to spiral reinforcement • Large amount of concrete inside spiral reinforcement missing • Some longitudinal bars ruptured • Shored	
11B,13B	• Missing	N/A
15B	• Spalled to spiral reinforcement from midheight to top • Shored	

COLUMN LABEL	DESCRIPTION OF DAMAGE	PHOTO (IF AVAILABLE)
7C	• Bowed • Spalled to spiral reinforcement • Some concrete missing inside spiral reinforcement • Shored	
9C	• Severely bowed • Stripped to spiral reinforcement • Lost some concrete inside spiral reinforcement • Shored	
11C	• Missing	N/A
13C	• Missing	N/A
15C	• Slightly blackened	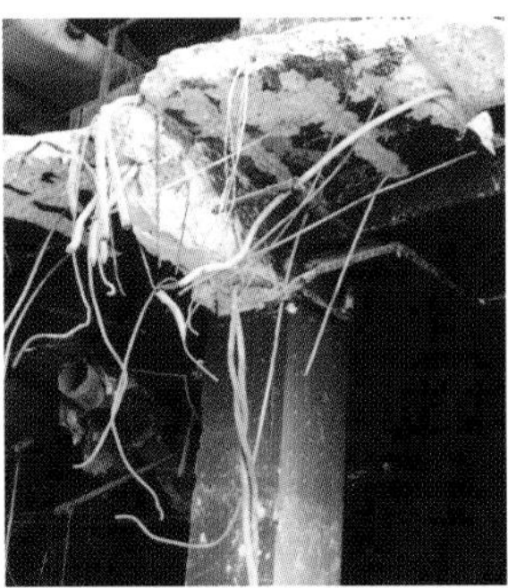
5D	• Intact • Minor edge spalling • Slightly blackened	

COLUMN LABEL	DESCRIPTION OF DAMAGE	PHOTO (IF AVAILABLE)
7D	• Split by vertical crack that looks typically like thermal cracking • Shored	
9D	• Missing • Shored	
11D	• Bowed • Stripped to spiral reinforcement • Shored	
13D	• Stripped to spiral reinforcement • Bowed to Northeast • Shored	

COLUMN LABEL	DESCRIPTION OF DAMAGE	PHOTO (IF AVAILABLE)
5E	• Downed • Stripped to spiral reinforcement • Shored	
7E,9E	• Missing • Shored	
11E	• Blackened	
3F	• Spalled to spiral reinforcement at top	

COLUMN LABEL	DESCRIPTION OF DAMAGE	PHOTO (IF AVAILABLE)
5F	• Broken at midheight • Remaining upper portion severely bent and stripped to spiral reinforcement • Concrete inside spiral broken • Shored	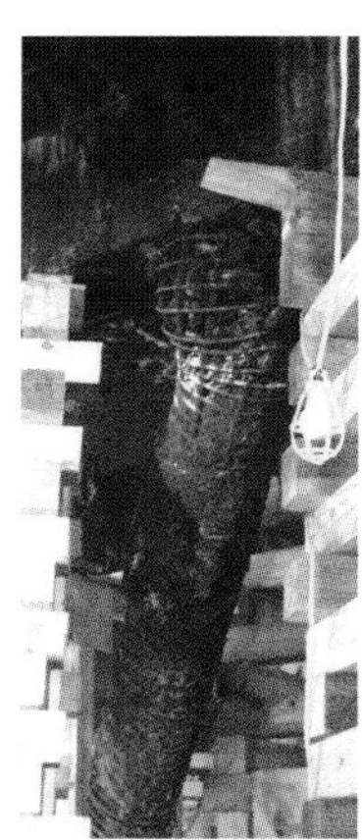
7F	• Missing	N/A
9F	• Bowed • Stripped to spiral reinforcement from floor to ~6 feet high • Shored	
11F	• Impacted, bowed • Stripped to spiral reinforcement • Shored	
3G	• Slightly blackened	

COLUMN LABEL	DESCRIPTION OF DAMAGE	PHOTO (IF AVAILABLE)
5G	• Disconnected at bottom • Severely bent • Only connected by steel at top • Shored	
7G	• Severe spalling • Steel visible midheight to top • Shored	N/A
9G	• Moderate spalling • Steel visible on northwest side at midsection	
1H-North	• Bowed • Stripped to spiral reinforcement • Shored	
1H-South	• Bowed • Stripped to spiral reinforcement from 3 feet above floor to top • Shored	

COLUMN LABEL	DESCRIPTION OF DAMAGE	PHOTO (IF AVAILABLE)
3H	• Bowed • Stripped to spiral reinforcement at midheight • Shored	
5H	• Downed • Stripped to spiral reinforcement • Shored	
7H	• Moderate spalling • Spalling more at top • No steel appears visible • Shored	
9H	• Moderate spalling • Some spiral reinforcement visible • Vertical cracks • Shored	

COLUMN LABEL	DESCRIPTION OF DAMAGE	PHOTO (IF AVAILABLE)
1J-North & 1J-South	• Intact • Blackened	
3J	• Bowed • Stripped to spiral reinforcement from 3 feet above floor to top • Shored	
5J	• Stripped to spiral reinforcement from 6 feet above floor to top • Shored	
7J	• Edge spall • Blackened • Light spalling	

COLUMN LABEL	DESCRIPTION OF DAMAGE	PHOTO (IF AVAILABLE)
1K-North	• Bowed to Northeast • Spiral reinforcement exposed at midheight • Shored	
1K-South	• Downed • Stripped to spiral reinforcement • Shored	
3K	• Missing • Shored	
5K	• Light spalling • Intact • Minor edge spalling	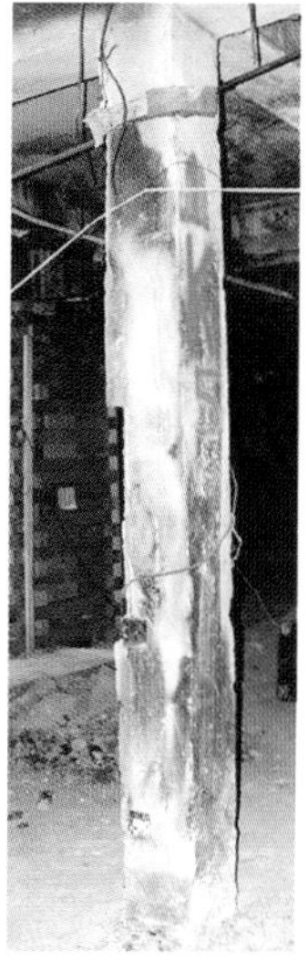

COLUMN LABEL	DESCRIPTION OF DAMAGE	PHOTO (IF AVAILABLE)
1L-North	• Intact • Slightly blackened	N/A
3L-North	• Damaged (per FBI)	N/A
1L-South	• Minor spalling • Intact • Slightly blackened	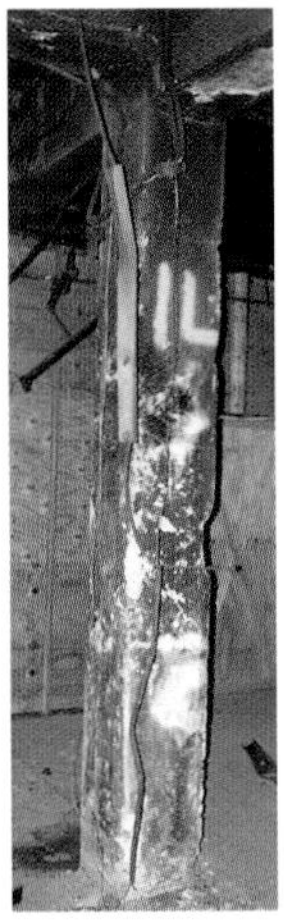
3L-South	• All spiral reinforcement exposed • Core concrete in place at bottom but missing at top • Deformation=30" • Shored	
5L-South	• Split • Stripped to spiral reinforcement at 4 feet above floor • Shored	
1M-North	• Damaged (per FBI)	N/A

COLUMN LABEL	DESCRIPTION OF DAMAGE	PHOTO (IF AVAILABLE)
3M-North	• Damaged (per FBI)	
5M-North	• Damaged (per FBI)	N/A
1M-South	• Spiral reinforcement fully exposed • Deformation is about 2–3 diameters • No cusps in shape • Bottom attached to slab, top attached by steel • Core concrete missing • Shored	
3M-South	• Edge spalling • Slightly blackened	
5M-South	• Edge spalling • Slightly blackened	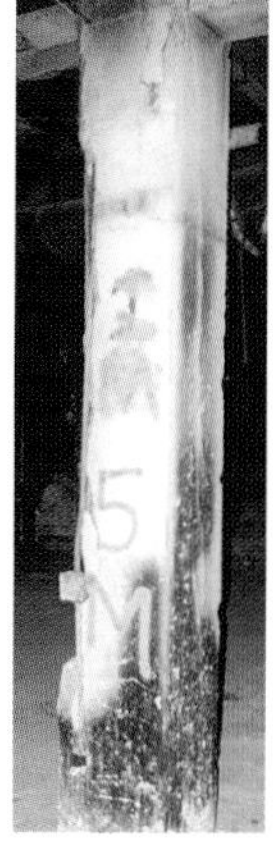

COLUMN LABEL	DESCRIPTION OF DAMAGE	PHOTO (IF AVAILABLE)
1N-North	• Damaged (per FBI)	N/A
3N-North	• Damaged (per FBI)	N/A
5N-North	• Damaged (per FBI)	N/A
1N-South	• Light spalling • Honeycombing (northeast corner)	
3N-South	• Moderate spalling at all corners • Bowed • Spiral steel exposed	
5N-South	• Moderate spalling • Concrete in place	

APPENDIX C

PHOTOGRAPHY SOURCES

Cover photo: AFP/Shawn Thew

Page 2: Pentagon Renovation Program Office

Page 3: Associated Press

Page 4: Department of Defense

Pages 14 and 15: Associated Press

Page 16, top: Associated Press

Page 16, bottom: Steve Riskus

Page 18: DefenseLink

Page 19, top: DefenseLink

Page 19, bottom: Air Survey Corporation

Page 24, both: BPS team

Page 25, top: DefenseLink

Page 25, bottom: BPS team

Page 26, all: BPS team

Page 27, all: BPS team

Page 28: BPS team

Page 29, both: BPS team

Page 30, top, both: BPS team

Page 30, bottom: Federal Bureau of Investigation

Page 31, all: BPS team

Page 32, all: BPS team

Page 33, all: BPS team

Page 40, both: BPS team

Page 41: BPS team

Page 42: BPS team

Page 43: BPS team

Page 44, all: BPS team

Page 47: BPS team

Page 48: BPS team

Page 51: Mike Johns

Appendix B, all: BPS team